AF736527

Studies in Choice and Welfare

Editor-in-Chief
M. Fleurbaey, USA
M. Salles, France

Series editors
B. Dutta, United Kingdom
W. Gaertner, Germany
C. Herrero Blanco, Spain
B. Klaus, Switzerland
P.K. Pattanaik, USA
K. Suzumura, Japan
W. Thomson, USA

More information about this series at http://www.springer.com/series/6869

Youngsub Chun

Fair Queueing

Springer

Youngsub Chun
Department of Economics
Seoul National University
Seoul
Republic of Korea

ISSN 1614-0311 ISSN 2197-8530 (electronic)
Studies in Choice and Welfare
ISBN 978-3-319-33770-8 ISBN 978-3-319-33771-5 (eBook)
DOI 10.1007/978-3-319-33771-5

Library of Congress Control Number: 2016940375

© Springer International Publishing Switzerland 2016
This work is subject to copyright. All rights are reserved by the Publisher, whether the whole or part of the material is concerned, specifically the rights of translation, reprinting, reuse of illustrations, recitation, broadcasting, reproduction on microfilms or in any other physical way, and transmission or information storage and retrieval, electronic adaptation, computer software, or by similar or dissimilar methodology now known or hereafter developed.
The use of general descriptive names, registered names, trademarks, service marks, etc. in this publication does not imply, even in the absence of a specific statement, that such names are exempt from the relevant protective laws and regulations and therefore free for general use.
The publisher, the authors and the editors are safe to assume that the advice and information in this book are believed to be true and accurate at the date of publication. Neither the publisher nor the authors or the editors give a warranty, express or implied, with respect to the material contained herein or for any errors or omissions that may have been made.

Printed on acid-free paper

This Springer imprint is published by Springer Nature
The registered company is Springer International Publishing AG Switzerland

To my parents

Acknowledgments

This work was supported by the National Research Foundation of Korea Grant funded by the Korean Government (KRF-2009-342-B00011).

In writing this book I have benefited in various ways from interactions with many people. First and foremost, I am deeply grateful to William Thomson for his excellent guidance and continuous encouragement throughout my academic life. This book would not be made possible without his help.

I thank my coauthors of various papers on the queueing problem, Eun Jeong Heo, Toru Hokari, Yuan Ju, Manipushpak Mitra, Suresh Mutuswami, René van den Brink, and Duygu Yengin for helping me to understand the subject better, and my students, Hee-In Chang, Yunji Her, and Hyukjun Kwon for their comments.

I am also grateful to two editors of this series, Maurice Salles and Marc Fleurbaey, for their comments, and Martina Bihn of Springer Verlag for her continuous encouragement on this project.

I would like to dedicate this book to my parents. Finally, I am extremely grateful to my wife, Jongim, and my daughter, Juhyun, for their personal support.

Contents

Chapter 1
Introduction

1.1 The Queueing Problem

Consider a group of agents who must be served in a facility. The facility can handle only one agent at a time and agents incur waiting costs. The *queueing problem* is concerned with finding the order in which to serve agents and the (positive or negative) monetary transfers they should receive. We assume that an agent's waiting cost is constant per unit of time, but that agents differ in their waiting costs. Each agent's utility is equal to the amount of her monetary transfer minus her total waiting cost. An *allocation* consists of each agent's position in the queue and the monetary transfer to her. An allocation is *feasible* if no two agents are assigned to the same position and the sum of transfers is not positive. An allocation rule, or simply a *rule,* associates with each problem a nonempty subset of feasible allocations.

A queueing problem arises when agents cannot coordinate on the time when they want to have a service (long queues at the grocery store, ATM machines, etc). Even if they can coordinate, all of them might have the same preferences, that is, all wants to have earlier service than later. The examples can easily be found in real life. Due to an ice storm, many business firms want to repair their electrical systems at the same time. All faculty members in the economics department want to move a new building at the same time. Many consumers want to implement a new computer program in their computers. Many researchers want to use a supercomputer or an expensive research facility.[1]

This problem can be solved by taking various approaches. First, the problem can be solved by applying solutions developed in the cooperative game theory (Chap. 3). Secondly, we can take an axiomatic approach. We propose a set of axioms which a desirable rule should satisfy and characterize all rules satisfying the set of axioms.

[1]For other interesting examples of the queueing problem, see Maniquet (2003), Mukherjee (2013), and Kayi and Ramaekers (2010).

© Springer International Publishing Switzerland 2016

Y. Chun, *Fair Queueing*, Studies in Choice and Welfare,
DOI 10.1007/978-3-319-33771-5_1

In addition to basic axioms, the main axiom can be motivated from a normative point of view as in Chaps. 4 and 5 or from a strategic point of view as in Chap. 6. In the latter chapter, we investigate the existence of rules satisfying *strategyproofness* which requires that an agent should not have an incentive to misrepresent her waiting cost no matter what she believes other agents. If strategyproof rules exist, then we can impose additional normative properties and check whether it is possible to satisfy all the properties as in Chaps. 6 and 7. Also, we can characterize the rules by imposing axioms relating the problem involving all agents to its subproblems with a smaller number of agents together with some normative or strategic axioms (Chap. 8). Finally, the problem can be solved by adopting a noncooperative approach which builds up a natural and intuitive bargaining protocol such that players can negotiate among themselves to resolve the queueing conflicts (Chap. 9).

The queueing problem can be generalized in many directions. However, we discuss only one possibility in Chap. 10. After generalizing the queueing problem by assuming that the facility has two parallel servers so that two agents can be served at the same time, we investigate whether the rules for the queueing problem with one server can be extended to this case and discuss their properties.

An earlier literature on the problem goes back to Dolan (1978) who provides a strategyproof, but not budget-balanced, rule for the problem. Later, Suijs (1996) and Mitra (2001, 2002) provide a strategyproof and budget-balanced rule for the problem. This rule is characterized by Kayi and Ramaekers (2010, in press), Hashimoto and Saitoh (2012), and Chun et al. (in press). On the other hand, Maniquet (2003) tries to solve the problem by applying a solution developed in the cooperative game theory and obtains the minimal transfer rule. He also shows that the same rule can be characterized by imposing axioms indicating how a rule should respond to changes in the queueing problem. This approach is further studied by Chun (2006a,b). Recently, there have been many papers which combine the strategic and the normative points of view. They characterize all rules satisfying *strategyproofness* together with some normative axioms (Kayi and Ramaekers 2010; Hashimoto and Saitoh 2012; Chun et al. 2014a,b, in press; Chun and Yengin 2014). Also, a noncooperative approach can be taken for the problem (Ju et al. 2014a,b). In this book, we study recent developments on the queueing problem.

1.2 Overview

Now we provide an overview of the content of each chapter. In Chap. 2, we introduce the basic concepts. First, we formally introduce the queueing problem and present basic axioms which a desirable rule should satisfy. Then, we define prominent rules for the problem and give a brief discussion on their properties.

In Chap. 3, we show that the queueing problem can be solved by applying solutions developed in cooperative game theory. To do so, queueing problems should be mapped into queueing games by defining a worth of coalition. We can define the worth of each coalition to be the minimum waiting cost incurred by its members under the optimistic assumption that they are served before the non-coalitional members. By applying the Shapley value (Shapley 1953) to the optimistic queueing game, we obtain the minimal transfer rule (Maniquet 2003). Alternatively, we can define the worth of each coalition to be the minimum total waiting cost incurred by its members under the pessimistic assumption that they are served after the non-coalitional members. If we apply the Shapley value to the pessimistic queueing game, we end up with a different rule, the maximal transfer rule (Chun 2006a). Next, we investigate what recommendations we have if other cooperative game theoretic solutions are applied to the queueing games. Surprisingly, we end up with the same recommendation: the Shapley value, the nucleolus (or the prenucleolus), and the τ-value coincide for queueing games (Chun and Hokari 2007).

In Chap. 4, we present characterizations of the minimal and the maximal transfer rules by imposing various axioms specifying how a rule should respond to changes in the waiting cost or population. Together with basic axioms, the minimal transfer rule is the only rule satisfying *independence of preceding costs,* or *negative cost monotonicity* and *last-agent equal responsibility* (Maniquet 2003), or *balanced consistency,* or *balanced cost reduction* (van den Brink and Chun 2012). On the other hand, the maximal transfer rule is the only rule satisfying *independence of following costs,* or *positive cost monotonicity* and *first-agent equal responsibility* (Chun 2006a), or *balanced consistency under constant completion time* (van den Brink and Chun 2012).

In Chap. 5, we explore the implications of *no-envy* (Foley 1967) in the context of queueing problems. *No-envy* requires that no agent should end up with a higher utility by consuming what any other agent consumes. First, it is not difficult to show that *no-envy* implies *queue-efficiency*. Then, we identify an easy way of checking whether a rule satisfies *no-envy.* The existence of such a rule can easily be established. We also ask whether there is a rule satisfying *efficiency* and *no-envy* together with either one of two cost monotonicity axioms, *negative cost monotonicity* and *positive cost monotonicity*. However, there is no rule satisfying *efficiency, no-envy,* and either one of two *cost monotonicity* axioms. To remedy the situation, we propose modifications of *no-envy, adjusted no-envy*, and *backward/forward no-envy.* Finally, we discuss whether three fairness requirements, *no-envy,* the *identical preferences lower bound,* and *egalitarian equivalence,* are compatible in this context. Chapter 5 is based mainly on Chun (2006b).

In Chap. 6, we study the implications of *strategyproofness* which requires that an agent should not have an incentive to misrepresent her waiting cost no matter what she believes other agents to be doing. We begin with the classic result of Holmström (1979) which implies in our context that a rule satisfies *queue-efficiency* and *strategyproofness* if and only if it is a VCG rule.[2] By additionally imposing *equal treatment of equals,* we characterize the complete family of anonymous VCG

[2]The family of VCG rules is due to Vickrey (1961), Clarke (1971), and Groves (1973).

rules. The symmetrically balanced VCG rule is the only member of this family satisfying *budget balance*. On the other hand, by imposing *independence* axioms, the pivotal and the reward-based pivotal rules (Mitra and Mutuswami 2011) can be characterized. We also characterize the class of *k*-pivotal rules by generalizing the independence axioms. Chapter 6 is based mainly on Chun et al. (2011, 2014b).

In Chap. 7, we investigate the implications of *egalitarian equivalence* (Pazner and Schmeidler 1978) together with *queue-efficiency* and *strategyproofness*. *Egalitarian equivalence* requires that for each problem, there should be a reference bundle such that each agent is indifferent between her bundle and the reference bundle. First, we provide a complete characterization of the family of rules satisfying the three axioms together. Although there is no rule in this family satisfying *budget balance, feasible* rules exist and we characterize the family of all such rules. We also show that it is impossible to find a rule satisfying *queue-efficiency, egalitarian equivalence,* and a stronger notion of strategyproofness, called *weak group strategyproofness*. This chapter is based mainly on Chun et al. (2014a).

In Chap. 8, we study the implications of *subgroup additivity* which requires that a rule assigns the same expected relative utility to each agent whether an agent's expected relative utility is calculated from the problem involving all agents or from its subproblems with a smaller number of agents. As a result, we present characterizations of five important rules: the minimal transfer rule, the maximal transfer rule, the pivotal rule, the reward-based pivotal rule, and the symmetrically balanced VCG rule. In addition to some basic axioms and *subgroup additivity,* the characterization results can be obtained by additionally imposing either a strategic axiom or an equity axiom. Chapter 8 is based mainly on Chun and Mitra (2014).

In Chap. 9, we investigate a strategic bargaining approach to resolve queueing conflicts. Given a situation where players with different waiting costs have to form a queue in order to be served, they firstly compete with each other for a specific position in the queue. The winner can decide to take up the position or sell it to the others. In the former case, the rest of the players proceed to compete for the remaining positions in the same manner, whereas in the latter case, the seller proposes a queue with corresponding payments to the others which can be accepted or rejected. Depending on which position players are going to compete for, the subgame perfect equilibrium outcome of the corresponding mechanism coincides with the payoff vector assigned by one of the two well-known rules for the queueing problem, either the maximal transfer rule or the minimal transfer rule, while an efficient queue is always formed in equilibrium. Chapter 9 is based mainly on Ju et al. (2014a,b).

Finally, in Chap. 10, we generalize the queueing problem by assuming that the facility has two parallel servers so that two agents can be served at the same time. Once again, we are interested in finding the order in which to serve agents and the monetary transfers they should receive. Similarly to the queueing problem with one server, we introduce the minimal transfer rule and the maximal transfer rule for the queueing problem with two parallel servers and show that they correspond to the Shapley (1953) value of queueing games with two parallel servers, for two alternative definitions of the worth of a coalition. If the worth of a coalition is defined by assuming the coalitional members are served before the non-coalitional members, then the minimal transfer rule is obtained. If it is defined by assuming the coalitional members are served after the non-coalitional members, then the maximal transfer rule is obtained. This chapter is based mainly on Chun and Heo (2008).

References

Chun, Y. (2006a). A pessimistic approach to the queueing problem. *Mathematical Social Sciences, 51*, 171–181.

Chun, Y. (2006b). No-envy in queueing problems. *Economic Theory, 29*, 151–162.

Chun, Y., & Heo, E. J. (2008). Queueing problems with two parallel servers. *International Journal of Economic Theory, 4*, 299–315.

Chun, Y., & Hokari. T. (2007). On the coincidence of the Shapley value and the nucleolus in queueing problems. *Seoul Journal of Economics, 20*(2), 223–237.

Chun, Y., & Mitra, M. (2014). Subgroup additivity in the queueing problem. *European Journal of Operational Research, 238*(1), 281–289.

Chun, Y., Mitra, M., & Mutuswami, S. (2011). No-envy, egalitarian equivalence and strategyproofness in queueing problems. Mimeo, Seoul National University.

Chun, Y., Mitra, M., & Mutuswami, S. (2014a). Egalitarian equivalence and strategyproofness in the queueing problem. *Economic Theory, 56*(2), 425–442.

Chun, Y., Mitra, M., & Mutuswami, S. (2014b). Characterizations of pivotal mechanisms in the queueing problem. *Mathematical Social Sciences, 72*, 62–66.

Chun, Y., Mitra, M., Mutuswami, S. (in press). A characterization of the symmetrically balanced VCG rule in the queueing problem. Games and Economic Behavior.

Chun, Y., & Yengin, D. (2014). Welfare lower bounds and strategyproofness in the queueing problem. Mimeo.

Clarke, E. H. (1971). Multi-part pricing of public goods. *Public Choice, 11*, 17–33.

Dolan, R. (1978). Incentive mechanisms for priority queueing problems. *Bell Journal of Economics, 9*, 421–436.

Foley, D. (1967). Resource allocation and the public sector. *Yale Economic Essays, 7*, 45–98.

Groves, T. (1973). Incentives in teams. *Econometrica, 41*, 617–631.

Hashimoto, K., & Saitoh, H. (2012). Strategy-proof and anonymous rule in queueing problems: a relationship between equity and efficiency. *Social Choice and Welfare, 38*, 473–480.

Holmström, B. (1979). Groves' schemes on restricted domains. *Econometrica, 47*, 1137–1144.

Ju, Y., Chun, Y., & van den Brink, R. (2014a). Auctioning and selling positions: A non-cooperative approach to queueing conflicts. Tinbergen Institute Discussion Paper TI 2014-016/II, The Netherlands.

Ju, Y., Chun, Y., & van den Brink, R. (2014b). Auctioning and selling positions: A non-cooperative approach to queueing conflicts. *Journal of Economic Theory, 153*, 33–45.

Kayi, C., & Ramaekers, E. (2010). Characterizations of Pareto-efficient, fair, and strategy-proof allocation rules in queueing problems. *Games and Economic Behavior, 68*, 220–232.

Kayi, C., & Ramaekers, E. (in press). Corrigendum to Characterizations of Pareto-efficient, fair, and strategy-proof allocation rules in queueing problems. [Games and Economic Behavior 68(2010):220–232]. Games and Economic Behavior.

Maniquet, F. (2003). A characterization of the Shapley value in queueing problems. *Journal of Economic Theory, 109*, 90–103.

Mitra, M. (2001). Mechanism design in queueing problems. *Economic Theory, 17*, 277–305.

Mitra, M. (2002). Achieving the first best in sequencing problems. *Review of Economic Design, 7*, 75–91.

Mitra, M., & Mutuswami, S. (2011). Group strategyproofness in queueing models. *Games and Economic Behavior, 72*, 242–254.

Mukherjee, C. (2013). Weak group strategy-proof and queue-efficient mechanisms for the queueing problem with multiple machines. *International Journal of Game Theory, 42*, 131–163.

Pazner, E., & Schmeidler, D. (1978). Egalitarian equivalent allocations: A new concept of equity. *Quarterly Journal of Economics, 92*, 671–687.

Shapley, L. S. (1953). A value for n-person games. In H. W. Kuhn & A. W. Tucker (Eds.), *Contributions to the theory of games II*. Annals of mathematics studies (pp. 307–317, Vol. 28). Princeton: Princeton University Press.

Suijs, J. (1996). On incentive compatibility and budget balancedness in public decision making. *Economic Design, 2*, 193–209.

van den Brink, R., & Chun, Y. (2012). Balanced consistency and balanced cost reduction for sequencing problems. *Social Choice and Welfare, 38*, 519–529.

Vickrey, W. (1961). Counterspeculation, auctions and competitive sealed tenders. *Journal of Finance, 16*, 8–37.

Chapter 2
Basic Concepts

2.1 Introduction

In this chapter, we introduce basic concepts which will be useful throughout the discussion. We begin with introducing the queueing problem which can be defined as the vector of unit waiting costs of agents. We are interested in finding the order in which to serve agents and the monetary transfers they should receive. We assume that an agent's waiting cost is constant per unit of time, but that agents differ in their waiting costs. An *allocation* consists of each agent's position in the queue and the monetary transfer to her. An allocation is *feasible* if no two agents are assigned to the same position and the sum of transfers is not positive. A *rule* associates with each problem a nonempty subset of feasible allocations.

To identify a well-behaved rule, we impose axioms which specify how a rule should make a recommendation in each queueing problem or how it should respond to certain changes in the queueing problem.[1] Our basic axioms can be divided into three efficiency and three fairness requirements. *Queue-efficiency* requires that the rule should choose queues which minimize the total waiting costs, *budget balance* requires that the sum of all transfers should be equal to zero, and *efficiency* (or *Pareto efficiency*) requires that the rule should satisfy both *queue-efficiency* and *budget balance*. On the other hand, *Pareto indifference* requires that a rule should choose all feasible allocations which give the same utilities to each agent, *equal treatment of equals* requires that two agents with the same waiting cost should end up with the same utilities, and the *identical preferences lower bound* (Moulin 1990) requires that each agent should be at least as well off as she would be, under *efficiency* and *equal treatment of equals*, if all other agents had the same preferences as her.

[1]As explained in Thomson (2010), the axioms can be organized into two main categories, punctual axioms and relational axioms. Punctual axioms specify how a rule should make a recommendation in each queueing problem whereas relational axioms specify how a rule should respond to certain changes in the queueing problem.

© Springer International Publishing Switzerland 2016
Y. Chun, *Fair Queueing*, Studies in Choice and Welfare,
DOI 10.1007/978-3-319-33771-5_2

For the queueing problem, many rules are proposed. Among them, most prominent ones are: the minimal transfer rule (Maniquet 2003), the maximal transfer rule (Chun 2006), the symmetrically balanced VCG rule, the pivotal rule, and the reward-based pivotal rule (Mitra and Mutuswami 2011). Both the minimum transfer and the maximum transfer rules are obtained by applying the Shapley value (Shapley 1953), one of the most widely discussed solution in cooperative games, to appropriately defined cooperative games of the queueing problem. For the minimal transfer rule, the worth of a coalition is defined to be the minimal waiting cost incurred by its members under the optimistic assumption that they are served before the non-coalitional members. For the maximal transfer rule, the worth of a coalition is defined to be the minimal waiting cost incurred by its members under the pessimistic assumption that they are served after the non-coalitional members. The next three rules satisfy *strategyproofness,* which requires truth telling to be a dominant strategy for each agent and for each (announced) state. The symmetrically balanced VCG rule is introduced by Suijs (1996) and Mitra (2001) and later characterized in the context of queueing problems by Kayi and Ramaekers (2010, in press) and Chun et al. (in press). The pivotal and the reward-based pivotal rules are introduced by Mitra and Mutuswami (2011) as two members of the family of k-pivotal rules which satisfy *weak group strategyproofness* requiring that any subgroup of agents cannot be made strictly better off by deviating.

This chapter is organized as follows. In Sect. 2.2, we formally introduce the queueing problem, and in Sect. 2.3, we present basic axioms which a desirable rule should satisfy. In Sect. 2.4, we define prominent rules for the problem and briefly discuss their properties.

2.2 The Model

Let $I \equiv \{1, 2, \ldots\}$ be an (infinite) universe of "potential" agents and $\mathcal{N}$ be the family of nonempty finite subsets of I. Each agent $i \in I$ is characterized by her unit waiting cost, $\theta_i \geq 0$. Given $N \in \mathcal{N}$, each agent $i \in N$ is assigned a position $\sigma_i \in \{1, \ldots, |N|\}$ in a queue[2] and a (positive or negative) transfer $t_i \in \mathbf{R}$. If the monetary transfer of an agent is positive, then this agent receives a compensation from other agents. If it is negative, she has to pay that amount as compensation to other agents. The agent who is served first incurs no waiting cost. Each agent has one job to process and the machine can process only one job at a time. Each job takes the same amount of processing time and without loss of generality, this processing time is normalized to one. If agent $i \in N$ is served in the σ_ith position, her waiting cost is $(\sigma_i - 1)\theta_i$.[3] An agent's net utility depends on her waiting costs

[2]For any set A, $|A|$ denotes the cardinality of A.

[3]This assumes that no waiting cost is incurred while the job is being processed. Alternatively, we can assume that if agent $i \in N$ is served in the σ_ith position, then her waiting cost is $\sigma_i \cdot \theta_i$. These

and the transfer she receives. Since we assume that each agent has a quasi-linear utility function, the utility of agent i from the bundle (σ_i, t_i) is given by

$$u(\sigma_i, t_i; \theta_i) = -(\sigma_i - 1)\theta_i + t_i.$$

For all $N \in \mathcal{N}$, let $\theta = (\theta_i)_{i\in N} \in \mathbf{R}^N_+$ be the profile of waiting costs of all agents.[4] For all profiles θ and all $i \in N$, let $\theta_{N\setminus\{i\}}$ be the profile of waiting costs of all agents except i. A *queueing problem* is defined as a profile $\theta \in \mathbf{R}^N_+$ where θ is the vector of unit waiting costs. Let $\mathcal{Q}^N$ be the class of all problems for N and $\mathcal{Q} = \cup\mathcal{Q}^N$. An *allocation* for $\theta \in \mathcal{Q}^N$ is a pair (σ, t), where for each $i \in N$, σ_i denotes agent i's position in the queue and t_i the monetary transfer to her. An allocation is *feasible* if no two agents are assigned to the same position and the sum of transfers is not positive. Thus, the set of feasible allocations $Z(\theta)$ consists of all pairs (σ, t) such that for all $i, j \in N$, $i \neq j$ implies $\sigma_i \neq \sigma_j$ and $\sum_{i\in N} t_i \leq 0$.

Given $\theta \in \mathcal{Q}^N$, an allocation $(\sigma, t) \in Z(\theta)$ is *queue-efficient* if it minimizes the total waiting costs, that is, for all $(\sigma', t') \in Z(\theta)$, $\sum_{i\in N}(\sigma_i - 1)\theta_i \leq \sum_{i\in N}(\sigma'_i - 1)\theta_i$. It is straightforward to check that an efficient queue serves agents in the non-increasing order of their waiting costs and that any queue with this property is also efficient. The efficient queue of a problem does not depend on the transfers. Moreover, it is unique except for agents with equal waiting costs. These agents have to be served consecutively but in any order. The set of efficient queues for $\theta \in \mathcal{Q}^N$ is denoted $\mathit{Eff}(\theta)$. An allocation $(\sigma, t) \in Z(\theta)$ is *budget balanced* if $\sum_{i\in N} t_i = 0$. An allocation rule, or simply a *rule*, is a mapping φ defined on $\mathcal{Q}$ which associates with every $N \in \mathcal{N}$ and every $\theta \in \mathcal{Q}^N$ a nonempty subset $\varphi(\theta)$ of feasible allocations. The pair $\varphi_i(\theta) = (\sigma_i, t_i)$ represents the position of agent i in the queue and her transfer in θ, which is a bundle assigned to agent i by φ. Given $\theta \in \mathcal{Q}^N$, $(\sigma, t) \in Z(\theta)$ and $i \in N$, let $P_i(\sigma) = \{j \in N \mid \sigma_j < \sigma_i\}$ be the set of agents preceding agent i in the queue σ and $F_i(\sigma) = \{j \in N \mid \sigma_j > \sigma_i\}$ the set of agents following her in the queue σ. When the context is clear, we abuse notation slightly by dropping the dependence on σ and simply referring to P_i and F_i. The set of all possible queues for N is $\Sigma(N)$. Similarly, for all $S \subseteq N$, the set of all possible queues for S is $\Sigma(S)$.

Remark 2.1 In Chaps. 6 and 7, to use the classic result of Holmström (1979), a rule is assumed to be single-valued and denoted by μ which associates to each $N \in \mathcal{N}$ and each $\theta \in \mathcal{Q}^N$, a tuple $\mu(\theta) = (\sigma, t) \in \Sigma(N) \times \mathbf{R}^n$. In these chapters, to choose a unique queue for each problem, we implicitly assume that there is an order of the agents which is used to break ties.

---

two definitions are equivalent in the following sense. Interchanging agents i and j in the queue leads to a saving of waiting cost in one definition if and only if it leads to a saving of waiting cost in the other definition. Formally, $\sigma_i\theta_i + \sigma_j\theta_j \leq \sigma'_i\theta_i + \sigma'_j\theta_j$ if and only if $(\sigma_i - 1)\theta_i + (\sigma_j - 1)\theta_j \leq (\sigma'_i - 1)\theta_i + (\sigma'_j - 1)\theta_j$.

[4]Here, $\mathbf{R}_+$ denotes the nonnegative orthant of the real line.

Remark 2.2 A queueing problem can be generalized to a *sequencing problem,* which is a list (r, θ), where $r \equiv (r_i)_{i \in N}$ is the vector representing the processing time of agents and $\theta \equiv (\theta_i)_{i \in N}$ is the vector of unit waiting costs. For a sequencing problem, each agent is characterized by the processing time and the unit waiting cost. A *queueing problem* is obtained by assuming that agents need the same amount of processing time, that is, for each $i \in N$, $r_i = 1$, but differ in the unit waiting cost. On the other hand, a *scheduling problem* is obtained by assuming that agents have the same unit waiting cost, that is, for each $i \in N$, $\theta_i = 1$, but differ in the amount of processing time. The sequencing problems[5] have been studied by Suijs (1996), van den Brink and Chun (2012), and others and the scheduling problem by Cres and Moulin (2001), Juarez (2008), Moulin (2007), and others.

2.3 Basic Axioms

Now we introduce basic axioms which we will impose on rules. *Queue-efficiency* requires that the rule should choose queues which minimize the total waiting costs. It is straightforward to check that an efficient queue serves agents in the nonincreasing order of their waiting costs, and any queue with this property is also efficient. *Budget balance* requires that the sum of all transfers should be equal to zero. It is a strengthening of *feasibility* which requires that the sum of all transfers should not be positive. *Efficiency* (or *Pareto efficiency*) requires that the rule should choose allocations that are *queue-efficient* and *budget balanced.*

Queue-efficiency: For all $N \in \mathcal{N}$, all $\theta \in \mathcal{Q}^N$, and all $(\sigma, t) \in \varphi(\theta)$, $\sigma \in Eff(\theta)$.
Budget balance: For all $N \in \mathcal{N}$, all $\theta \in \mathcal{Q}^N$, and all $(\sigma, t) \in \varphi(\theta)$, $\sum_{i \in N} t_i = 0$.
Efficiency: For all $N \in \mathcal{N}$, all $\theta \in \mathcal{Q}^N$, and all $(\sigma, t) \in \varphi(\theta)$, $\sigma \in Eff(\theta)$ and $\sum_{i \in N} t_i = 0$.

Pareto indifference requires that if an allocation is chosen by a rule, then all other feasible allocations which assign the same utilities to each agent should be chosen by the rule. It requires for a rule to choose all feasible allocations which give the same utilities to each agent. *Equal treatment of equals* requires that two agents with the same waiting cost should end up with the same utilities.

Pareto indifference: For all $N \in \mathcal{N}$, all $\theta \in \mathcal{Q}^N$, all $(\sigma, t) \in \varphi(\theta)$, and $(\sigma', t') \in Z(\theta)$, if for all $i \in N$, $u(\sigma'_i, t'_i; \theta_i) = u(\sigma_i, t_i; \theta_i)$, then $(\sigma', t') \in \varphi(\theta)$.
Equal treatment of equals: For all $N \in \mathcal{N}$, all $\theta \in \mathcal{Q}^N$, all $(\sigma, t) \in \varphi(\theta)$, and all $i, j \in N$, if $\theta_i = \theta_j$, then $u(\sigma_i, t_i; \theta_i) = u(\sigma_j, t_j; \theta_j)$.

Finally, the *identical preferences lower bound* (Moulin 1990) requires that each agent should be at least as well off as she would be, under *efficiency* and *equal*

[5] Also, see Curiel et al. (1989) for a sequencing problem with an initial queue and Chun (2011) for a sequencing problem with bilateral transfers.

treatment of equals, if all other agents had the same preferences as her. Note that if a rule satisfies *efficiency* and *equal treatment of equals* and all agents have the same waiting costs as agent i, then all agents end up with the same utilities of $-\frac{|N|-1}{2}\theta_i$. The *identical preferences lower bound* requires that all agents should be better off by not having the same waiting cost.

Identical preferences lower bound: For all $N \in \mathcal{N}$, all $\theta \in \mathcal{Q}^N$, all $(\sigma, t) \in \varphi(\theta)$, and all $i \in N$, $u(\sigma_i, t_i; \theta_i) \geq -\frac{|N|-1}{2}\theta_i$.

2.4 Rules

In this section, we introduce important rules for the queueing problem. We first consider a two-agent queueing problem. Suppose that there are two agents denoted by agents 1 and 2 and that $\sigma_1 < \sigma_2$. If agent 2 moves up, then her utility gains are θ_2. She enjoys the same utility whether she receives $\frac{\theta_2}{2}$ at σ_2 or pays the same amount at σ_1. On the other hand, if agent 1 is served later, then her utility losses are θ_1. She enjoys the same utility whether she pays $\frac{\theta_1}{2}$ at σ_1 or receives the same amount at σ_2. Therefore, it is natural to expect the actual transfer will be determined by these two bounds. The following two rules select an efficient queue and transfer either the minimum or the maximum of these two bounds for two-agent problems.

The minimal transfer rule (Maniquet 2003), which chooses the minimum of the two bounds for two-agent problems, selects an efficient queue and transfers to each agent a half of her unit waiting cost multiplied by the number of her predecessors minus a half of the sum of the unit waiting costs of her followers.

Minimal transfer rule, φ^M: For all $N \in \mathcal{N}$ and all $\theta \in \mathcal{Q}^N$,

$$\varphi^M(\theta) = \{(\sigma^M, t^M) \in Z(\theta) |\ \sigma^M \in \mathit{Eff}(\theta) \text{ and } \forall i \in N,$$

$$t_i^M = (\sigma_i^M - 1)\frac{\theta_i}{2} - \sum_{j \in F_i(\sigma^M)} \frac{\theta_j}{2}\}.$$

On the other hand, the maximal transfer rule (Chun 2006), which chooses the maximum of the two bounds for two-agent problems, selects an efficient queue and transfers to each agent a half of the sum of the unit waiting costs of her predecessors minus a half of her unit waiting cost multiplied by the number of her followers.

Maximal transfer rule, φ^C: For all $N \in \mathcal{N}$ and all $\theta \in \mathcal{Q}^N$,

$$\varphi^C(\theta) = \{(\sigma^C, t^C) \in Z(\theta) |\ \sigma^C \in \mathit{Eff}(\theta) \text{ and } \forall i \in N,$$

$$t_i^C = \sum_{j \in P_i(\sigma^C)} \frac{\theta_j}{2} - (|N| - \sigma_i^C)\frac{\theta_i}{2}\}.$$

Both the minimum and the maximum transfer rules satisfy *queue-efficiency, budget balance, efficiency, Pareto indifference, equal treatment of equals,* and the *identical preferences lower bound.* Moreover, these two rules are obtained by applying the Shapley value (Shapley 1953), one of the most widely discussed solution in cooperative games, to appropriately define cooperative games of the queueing problem. For the minimal transfer rule, the worth of a coalition is defined to be the minimal waiting cost incurred by its members under the optimistic assumption that they are served before the non-coalitional members. For the maximal transfer rule, the worth of a coalition is defined to be the minimal waiting cost incurred by its members under the pessimistic assumption that they are served after the non-coalitional members (see Chap. 3 for details).

We note that the minimal and the maximal transfer rules assign a unique allocation if and only if all agents have different waiting costs. If two agents have the same waiting cost, then the efficient queue is not unique, and consequently the allocations chosen by the rule are not unique either. However, agents' utilities do not depend on the choice of efficient queues if the transfer is determined according to the minimal or the maximal transfer rule. Thus, both rules are *essentially single-valued,* in the sense that for a given problem, each agent's utility is the same at all allocations that the rule chooses. As a consequence, any efficient queue can be chosen to calculate the utilities assigned by the two rules. To be specific, for all $N \in \mathcal{N}$ and all $\theta \in \mathcal{Q}^N$, for the minimal transfer rule, the utility of agent i is given by

$$\begin{aligned} u_i(\sigma^M, t^M) &= -(\sigma_i^M - 1)\theta_i + t_i^M \\ &= -(\sigma_i^M - 1)\theta_i + (\sigma_i^M - 1)\frac{\theta_i}{2} - \sum_{j \in F_i(\sigma^M)} \frac{\theta_j}{2} \\ &= -(\sigma_i^M - 1)\frac{\theta_i}{2} - \sum_{j \in F_i(\sigma^M)} \frac{\theta_j}{2}, \end{aligned}$$

and for the maximal transfer rule,

$$\begin{aligned} u_i(\sigma^C, t^C) &= -(\sigma_i^C - 1)\theta_i + t_i^C \\ &= -(\sigma_i^C - 1)\theta_i + \sum_{j \in P_i(\sigma^C)} \frac{\theta_j}{2} - (n - \sigma_i^C)\frac{\theta_i}{2} \\ &= -(n - 1)\theta_i + \sum_{j \in P_i(\sigma^C)} \frac{\theta_j}{2} + (n - \sigma_i^C)\frac{\theta_i}{2}. \end{aligned}$$

The next three rules satisfy *strategyproofness,* which requires truth telling to be a dominant strategy for each agent and for each (announced) state. The first rule, which can be called the symmetrically balanced VCG rule,[6] was introduced by Suijs (1996) and Mitra (2001) and later characterized in the context of queueing problems by Kayi and Ramaekers (2010, in press) and Chun et al. (in press).

Symmetrically balanced VCG rule, φ^B: For all $N \in \mathcal{N}$ with $|N| \geq 3$ and all $\theta \in \mathcal{Q}^N$,

$$\varphi^B(\theta) = \{(\sigma^B, t^B) \in Z(\theta) | \ \sigma^B \in \mathit{Eff}(\theta), \ \text{and} \ \forall i \in N,$$

$$t_i^B = \sum_{j \in P_i(\sigma^B)} \left(\frac{\sigma_j^B - 1}{|N| - 2}\right)\theta_j - \sum_{k \in F_i(\sigma^B)} \left(\frac{|N| - \sigma_k^B}{|N| - 2}\right)\theta_k\}. \tag{2.1}$$

The symmetrically balanced VCG rule satisfies *queue-efficiency, budget balance, efficiency, Pareto indifference, equal treatment of equals,* and the *identical preferences lower bound.* Moreover, it is the only rule satisfying *queue-efficiency, budget balance, Pareto indifference, equal treatment of equals*, and *strategyproofness* (see Sect. 6.3.1 for details). The queueing problem is one of rare problems in which the five requirements can be satisfied together.

Mitra and Mutuswami (2011) introduce and characterize the family of k-pivotal rules on the basis of *pairwise strategyproofness,* which requires that as long as there is no further side payments across agents, there does not exist any pair of agents that can benefit by deviating from truth telling. Moreover, all k-pivotal rules satisfy *weak group strategyproofness,* which requires that any subgroup of agents cannot be made strictly better off by deviating. The pivotal and the reward-based pivotal rules belong to the family of k-pivotal rules (see Sect. 6.3.2 for details).

Pivotal rule, φ^P: For all $N \in \mathcal{N}$ and all $\theta \in \mathcal{Q}^N$,

$$\varphi^P(\theta) = \{(\sigma^P, t^P) \in Z(\theta) | \ \sigma^P \in \mathit{Eff}(\theta) \ \text{and} \ \forall i \in N, \ t_i^P = -\sum_{j \in F_i(\sigma^P)} \theta_j\}.$$

Reward-based pivotal rule, φ^R: For all $N \in \mathcal{N}$ and all $\theta \in \mathcal{Q}^N$,

$$\varphi^R(\theta) = \{(\sigma^R, t^R) \in Z(\theta) | \ \sigma^R \in \mathit{Eff}(\theta) \ \text{and} \ \forall i \in N, \ t_i^R = \sum_{j \in P_i(\sigma^R)} \theta_j\}.$$

Both the pivotal and the reward-based pivotal rules satisfy *queue-efficiency, Pareto indifference,* and *equal treatment of equals,* but fail to satisfy *budget balance* and *efficiency*. The pivotal rule does not satisfy the *identical preferences lower bound,* but the reward-based pivotal rule does.

[6]The family of VCG rules is due to Vickrey (1961), Clarke (1971), and Groves (1973).

References

Chun, Y. (2006). A pessimistic approach to the queueing problem. *Mathematical Social Sciences, 51*, 171–181.

Chun, Y. (2011). Consistency and monotonicity in sequencing problems. *International Journal of Game Theory, 40*, 29–41.

Chun, Y., Mitra, M., & Mutuswami, S. (in press). A characterization of the symmetrically balanced VCG rule in the queueing problem. Games and Economic Behavior.

Clarke, E. H. (1971). Multi-part pricing of public goods. *Public Choice, 11*, 17–33.

Cres, H., & Moulin, H. (2001). Scheduling with opting out: Improving upon random priority. *Operations Research, 49*, 565–577.

Curiel, I., Pederzoli, G., & Tijs, S. (1989). Sequencing games. *European Journal of Operations Research, 40*, 344–351.

Groves, T. (1973). Incentives in teams. *Econometrica, 41*, 617–631.

Holmström, B. (1979). Groves' schemes on restricted domains. *Econometrica, 47*, 1137–1144.

Juarez, R. (2008). The worst absolute surplus loss in the problem of commons: Random priority versus average cost. *Economic Theory, 34*, 69–84.

Kayi, C., & Ramaekers, E. (2010). Characterizations of Pareto-efficient, fair, and strategy-proof allocation rules in queueing problems. *Games and Economic Behavior, 68*, 220–232.

Kayi, C., & Ramaekers, E. (in press). Corrigendum to Characterizations of Pareto-efficient, fair, and strategy-proof allocation rules in queueing problems [Games and Economic Behavior 68:220–232]. Games and Economic Behavior.

Maniquet, F. (2003). A characterization of the Shapley value in queueing problems. *Journal of Economic Theory, 109*, 90–103.

Mitra, M. (2001). Mechanism design in queueing problems. *Economic Theory, 17*, 277–305.

Mitra, M., & Mutuswami, S. (2011). Group strategyproofness in queueing models. *Games and Economic Behavior, 72*, 242–254.

Moulin, H. (1990). Fair division under joint ownership: Recent results and open problems. *Social Choice and Welfare, 7*, 149–170.

Moulin, H. (2007). On scheduling fees to prevent merging, splitting, and transferring of jobs. *Mathematics of Operations Research, 32*, 266–283.

Shapley, L. S. (1953). A Value for n-Person Games. In H. W. Kuhn & A. W. Tucker (Eds.), *Contributions to the theory of games II*. Annals of mathematics studies (pp. 307–317, Vol. 28). Princeton: Princeton University Press.

Suijs, J. (1996). On incentive compatibility and budget balancedness in public decision making. *Economic Design, 2*, 193–209.

Thomson. W. (2010). Fair allocation rules. Mimeo.

van den Brink, R., & Chun, Y. (2012). Balanced consistency and balanced cost reduction for sequencing problems. *Social Choice and Welfare, 38*, 519–529.

Vickrey, W. (1961). Counterspeculation, auctions and competitive sealed tenders. *Journal of Finance, 16*, 8–37.

Chapter 3
Cooperative Game Theoretic Approach

3.1 Introduction

The *queueing problem* can be solved by applying solutions developed in cooperative game theory. To do this, queueing problems should be mapped into queueing games by defining a worth of coalition. We can define the worth of a coalition to be the minimum waiting cost incurred by its members under the optimistic assumption that they are served before the non-coalitional members (Maniquet 2003). By applying what is probably the best-known cooperative game theoretic solution, the Shapley value (Shapley 1953), to the optimistic queueing game, we obtain the minimal transfer rule which selects an efficient queue and transfers to each agent a half of her unit waiting cost multiplied by the number of her predecessors minus a half of the sum of the unit waiting costs of her followers.

Alternatively, we can take a pessimistic approach and define the worth of each coalition to be the minimum waiting cost incurred by its members when they are served after the non-coalitional members (Chun 2006). By applying the Shapley value to the pessimistic queueing game, we obtain a different rule, the maximal transfer rule, which selects an efficient queue and transfers to each agent a half of the sum of the unit waiting costs of her predecessors minus a half of her unit waiting cost multiplied by the number of her followers.

These results show the importance of the definition of the worth of a coalition in queueing problems. For some classes of problems,[1] it makes no difference whether the coalitional members have priority over the non-coalitional members or the non-coalitional members have priority over the coalitional members. If the Shapley value is applied, we obtain the same recommendation. However, for queueing problems, this is not the case: Depending upon who has priority, the resulting rule has very different properties.

[1]For example, the bankruptcy problem discussed in Remark 3.1.

© Springer International Publishing Switzerland 2016

Y. Chun, *Fair Queueing*, Studies in Choice and Welfare,
DOI 10.1007/978-3-319-33771-5_3

Next, we apply another well-known solution of cooperative games, the nucleolus (Schmeidler 1969) (or the prenucleolus) to queueing games, and identify the resulting rule. Surprisingly, we obtain the same rule: The Shapley value and the nucleolus (or the prenucleolus) coincide for queueing games. Also, we investigate the relation between the minimal and the maximal transfer rules and other rules discussed in the literature, the core, the τ-value (Tijs 1987), and the serial cost sharing rule (Moulin and Shenker 1992).

The chapter is organized as follows.[2] In Sect. 3.2, we introduce the basic concepts in cooperative game theory. In Sect. 3.3, we introduce the optimistic queueing game and show that the Shapley value applied to the resulting game corresponds to the minimal transfer rule. In Sect. 3.4, we introduce the pessimistic queueing game and show that the Shapley value applied to the resulting game corresponds to the maximal transfer rule. In Sect. 3.5, we establish the coincidence between the Shapley value and the nucleolus (or the prenucleolus) in the queueing games. In Sect. 3.6, we investigate the relation between the minimal and the maximal transfer rules and other rules discussed in the literature, the core, the τ-value, and the serial cost sharing rule. Concluding remarks are in Sect. 3.7.

3.2 Cooperative Games

We formally describe cooperative games with transferable utility or *games*. Let $N \equiv \{1, 2, \ldots, n\}$ be the set of *players*. A set $S \subseteq N$ is a *coalition*. A game is a real-valued function v defined on all coalitions $S \subseteq N$ satisfying $v(\emptyset) = 0$. The number $v(S)$ is the *worth* of S. Let Γ^N be the class of games with player set N. A *solution* is a function $\phi\colon \Gamma^N \to \mathbf{R}^N$, which associates with every game $v \in \Gamma^N$ a vector $\phi(v) = (\phi_i(v))_{i \in N} \in \mathbf{R}^N$. The number $\phi_i(v)$ represents the payoff to player i in game v.

Now we introduce two well-known solutions for games, the Shapley value (Shapley 1953) and the nucleolus (Schmeidler 1969). The Shapley value assigns to each player a payoff equal to a weighted average of her marginal contributions to all possible coalitions, with weights being determined by the sizes of coalitions. The nucleolus chooses the unique allocation from the set of imputations which minimizes the excess of coalitions in the lexicographic way. On the other hand, the prenucleolus chooses the unique allocation from the set of efficient allocations which minimizes the excess of the coalitions in the lexicographic way.

[2]The results of this chapter are collected from Maniquet (2003), Chun (2006), and Chun and Hokari (2007). Excerpts from Chun (2006) are reprinted with kind permission of Elsevier. Excerpts from Chun and Hokari (2007) are reprinted with kind permission of Institute of Economic Research, Seoul National University.

Shapley value, ***Sh*****:** For all $v \in \Gamma^N$ and all $i \in N$,

$$Sh_i(v) = \sum_{S \subseteq N, S \ni i} \frac{(|S|-1)!|N \backslash S|!}{|N|!} \big[v(S) - v(S \setminus \{i\})\big].$$

For all $v \in \Gamma^N$, let $X(v) = \{x \in \mathbf{R}^N | \sum_{i \in N} x_i = v(N)\}$ be the set of efficient allocations for v and $I(v) = \{x \in \mathbf{R}^N | \sum_{i \in N} x_i = v(N)$ and for all $i \in N$, $x_i \geq v(\{i\})\}$ be the set of *imputations* for v. For all $x \in I(v)$, its *excess vector* $e(v,x) \in \mathbf{R}^{2^N}$ is defined by setting for all $S \subseteq N$, $e_S(v,x) \equiv v(S) - \sum_{i \in S} x_i$. Its S-coordinate $e_S(v,x)$ measures the amount by which the worth of the coalition S exceeds its payoff at x. For all $y \in \mathbf{R}^{2^N}$, let $\tilde{y} \in \mathbf{R}^{2^{|N|}}$ be obtained by rearranging the coordinates of y in nonincreasing order. For all $y, z \in \mathbf{R}^{2^N}$, *$y$ is lexicographically smaller than z* if either (1) $\tilde{y}_1 < \tilde{z}_1$ or (2) there exists $\ell > 1$ such that $\tilde{y}_\ell < \tilde{z}_\ell$ and for all $k < \ell$, $\tilde{y}_k = \tilde{z}_k$.

Nucleolus, ***Nu*****:** For all $v \in \Gamma^N$ such that $I(v) \neq \emptyset$,

$$Nu(v) \equiv \left\{ x \in I(v) \,\middle|\, \begin{array}{l} \text{for all } x' \in I(v) \backslash \{x\},\ e(v,x) \text{ is} \\ \text{lexicographically smaller than } e(v,x') \end{array} \right\}$$

Prenucleolus, ***PN*****:** For all $v \in \Gamma^N$,

$$PN(v) \equiv \left\{ x \in X(v) \,\middle|\, \begin{array}{l} \text{for all } x' \in X(v) \backslash \{x\},\ e(v,x) \text{ is} \\ \text{lexicographically smaller than } e(v,x') \end{array} \right\}$$

For all $v \in \Gamma^N$, the *core* is the set of imputations at which no excess is greater than zero, that is, $Core(v) \equiv \{x \in I(v) |$ for all $S \subset N$, $\sum_{i \in S} x_i \geq v(S)\}$. A game is *convex* if for all $S, T \subseteq N$, $v(S) + v(T) \leq v(S \cup T) + v(S \cap T)$. It is well-known that a convex game has a nonempty core. Moreover, the Shapley value and the nucleolus select allocations in the core.

3.3 An Optimistic Approach and the Minimal Transfer Rule

We analyze queueing problems by applying solutions of games after defining a worth of coalition. We can define the worth of each coalition $S \subseteq N$ to be the minimum waiting cost incurred by its members under the optimistic assumption that they are served before the non-coalitional members. That is, for all $S \subseteq N$, its worth $v_O(S)$ of the optimistic queueing game is defined by setting:

$$v_O(S) = -\sum_{i \in S} (\sigma_i^* - 1)\theta_i,$$

where $\theta_S = (\theta_i)_{i \in S}$ and $\sigma^* \in Eff(\theta_S)$. By applying the Shapley value to the optimistic queueing game $v_O = (v_O(S))_{S \subseteq N}$, we show that the resulting payoff to each player is equal to the utility assigned by the minimal transfer rule.

Theorem 3.1 (Maniquet 2003) *Let $\theta \in \mathcal{Q}^N$. Let $z = (\sigma, t) \in Z(\theta)$ be such that agents' utilities at z are equal to the payoff vector obtained by applying the Shapley value to v_O. Then, $\sigma \in Eff(\theta)$ and for all $i \in N$,*

$$t_i = (\sigma_i - 1)\frac{\theta_i}{2} - \sum_{j \in F_i(\sigma)} \frac{\theta_j}{2}.$$

Proof It is well-known that a game can be expressed as a sum of unanimity games, that is, $v = \sum_{T \subseteq N} \lambda_v(T) u_T$, where the unanimity game u_T is defined by $u_T(S) = 1$ if $T \subseteq S$, and $u_T(S) = 0$ otherwise. For all $S \subseteq N$, the dividend $\lambda_v(S)$ is defined as follows: if $|S| = 1$, $\lambda_v(S) = v(S)$, and if $|S| > 1$, $\lambda_v(S) = v(S) - \sum_{T \subset S, T \neq S} \lambda_v(T)$.

We claim that for all $\theta \in \mathcal{Q}^N$,

$$\lambda_{v_O}(S) = \begin{cases} 0 & \text{if } S = \{i\}, \\ -\min_{i \in S} \theta_i & \text{if } |S| = 2, \\ 0 & \text{if } |S| \geq 3. \end{cases} \tag{3.1}$$

Indeed, if $S = \{i\}$, then $\lambda_{v_O}(S) = v_O(S) = 0$. For $|S| = 2$, we assume without loss of generality that $S = \{i, j\}$ and $\theta_i \geq \theta_j$. Then, $\lambda_{v_O}(S) = v_O(S) - \lambda_{v_O}(\{i\}) - \lambda_{v_O}(\{j\}) = -\theta_j$, as desired. For $|S| = 3$, we assume without loss of generality that $S = \{i, j, k\}$ and $\theta_i \geq \theta_j \geq \theta_k$. Then, $\lambda_{v_O}(S) = v_O(S) - \lambda_{v_O}(\{i, j\}) - \lambda_{v_O}(\{j, k\}) - \lambda_{v_O}(\{i, k\}) - \lambda_{v_O}(\{i\}) - \lambda_{v_O}(\{j\}) - \lambda_{v_O}(\{k\}) = -\theta_j - 2\theta_k + \theta_j + \theta_k + \theta_k = 0$, as desired. Let $S \subseteq N$ be such that $|S| > 3$. We assume without loss of generality that $S = \{1, 2, \ldots, s\}$ and $\theta_i \geq \theta_j$ for each $i \leq j$. As induction hypothesis, suppose that $\lambda_{v_O}(S') = 0$ for all $3 \leq |S'| \leq |S|$. Then, $\lambda_{v_O}(S) = v_O(S) - \sum_{T \subset S, |T| = 2} \lambda_{v_O}(T) = -\sum_{h=1}^{s} (\sigma_h - 1)\theta_h - \sum_{h=1}^{s} (-(\sigma_h - 1))\theta_h = 0$, as desired.

On the other hand, the Shapley value of player $i \in N$ in game v is given by $Sh_i(v) = \sum_{S \subseteq N, i \in S} \frac{\lambda_v(S)}{|S|}$. By substituting Eq. (3.1) into this expression, we obtain

$$Sh_i(v_O) = -(\sigma_i - 1)\frac{\theta_i}{2} - \sum_{j \in F_i(\sigma)} \frac{\theta_j}{2},$$

where $\sigma \in Eff(\theta)$. Using $t_i = u(\sigma_i, t_i; \theta_i) + (\sigma_i - 1)\theta_i$, we have

$$\begin{aligned} t_i &= -(\sigma_i - 1)\frac{\theta_i}{2} - \sum_{j \in F_i(\sigma)} \frac{\theta_j}{2} + (\sigma_i - 1)\theta_i \\ &= (\sigma_i - 1)\frac{\theta_i}{2} - \sum_{j \in F_i(\sigma)} \frac{\theta_j}{2}, \end{aligned}$$

the desired conclusion. □

3.4 A Pessimistic Approach and the Maximal Transfer Rule

We propose an alternative definition of the worth of a coalition which results in a very different rule, even if the same Shapley value is applied. By taking a pessimistic approach, we assume that for each $S \subseteq N$, the members of S are served after the members of $N \backslash S$. Since the members of S need to wait $|N| - |S|$ time periods to be served, its worth, $v_P(S)$, is defined by setting:

$$v_P(S) = -\sum_{i \in S}(|N| - |S| + \sigma_i^* - 1)\theta_i,$$

where $\theta_S = (\theta_i)_{i \in S}$ and $\sigma^* \in Eff(\theta_S)$. Now, we apply the Shapley value to the pessimistic queueing game $v_P = (v_P(S))_{S \subseteq N}$, and show that the resulting payoff to each player is equal to the utility assigned by the maximal transfer rule.

Theorem 3.2 (Chun 2006) *Let $\theta \in \mathcal{Q}^N$. Let $z = (\sigma, t) \in Z(\theta)$ be such that agents' utilities at z are equal to the payoff vector obtained by applying the Shapley value to v_P. Then, $\sigma \in Eff(\theta)$ and for all $i \in N$,*

$$t_i = \sum_{j \in P_i(\sigma)} \frac{\theta_j}{2} - (|N| - \sigma_i)\frac{\theta_i}{2}.$$

Proof As in the beginning of the proof of Theorem 3.1, a game can be expressed as a sum of unanimity games and the dividend is defined in the same way. We claim that for all $\theta \in \mathcal{Q}^N$,

$$\lambda_{v_P}(S) = \begin{cases} -(|N| - 1)\theta_i & \text{if } S = \{i\}, \\ \max_{i \in S} \theta_i & \text{if } |S| = 2, \\ 0 & \text{if } |S| \geq 3. \end{cases} \tag{3.2}$$

Indeed, if $S = \{i\}$, then $\lambda_{v_P}(S) = v_P(S) = -(|N| - 1)\theta_i$. For $|S| = 2$, we assume without loss of generality that $S = \{i, j\}$ and $\theta_i \geq \theta_j$. Then, $\lambda_{v_P}(S) = v_P(S) - \lambda_{v_P}(\{i\}) - \lambda_{v_P}(\{j\}) = -(|N| - 2)\theta_i - (|N| - 1)\theta_j + (|N| - 1)\theta_i + (|N| - 1)\theta_j = \theta_i$, as desired. For $|S| = 3$, we assume without loss of generality that $S = \{i, j, k\}$ and $\theta_i \geq \theta_j \geq \theta_k$. Then, $\lambda_{v_P}(S) = v_P(S) - \lambda_{v_P}(\{i, j\}) - \lambda_{v_P}(\{j, k\}) - \lambda_{v_P}(\{i, k\}) - \lambda_{v_P}(\{i\}) - \lambda_{v_P}(\{j\}) - \lambda_{v_P}(\{k\}) = -(|N| - 3)\theta_i - (|N| - 2)\theta_j - (|N| - 1)\theta_k - \theta_i - \theta_j - \theta_i + (|N| - 1)\theta_i + (|N| - 1)\theta_j + (|N| - 1)\theta_k = 0$, as desired. Let $S \subseteq N$ be such that $|S| > 3$. We assume without loss of generality that $S = \{1, 2, \ldots, s\}$ and $\theta_i \geq \theta_j$ for each $i \leq j$. As induction hypothesis, suppose that $\lambda_{v_P}(S') = 0$ for all $3 \leq |S'| \leq |S|$. Then, $\lambda_{v_P}(S) = v_P(S) - \sum_{T \subset S, |T|=1,2} \lambda_{v_P}(T) = -\sum_{h=1}^{s}(|N| - |S| + \sigma_h - 1)\theta_h - \sum_{h=1}^{s}(|S| - \sigma_h)\theta_h + \sum_{h=1}^{s}(|N| - 1)\theta_h = 0$, as desired.

On the other hand, the Shapley value of player $i \in N$ in game v is given by $Sh_i(v) = \sum_{S \subseteq N, i \in S} \frac{\lambda_v(S)}{|S|}$. By substituting Eq. (3.2) into this expression, we obtain

$$Sh_i(v_P) = -(|N|-1)\theta_i + \{ \sum_{j \in P_i(\sigma)} \frac{\theta_j}{2} + (|N|-\sigma_i)\frac{\theta_i}{2} \},$$

where $\sigma \in Eff(\theta)$. Using $t_i = u(\sigma_i, t_i; \theta_i) + (\sigma_i - 1)\theta_i$, we have

$$\begin{aligned} t_i &= -(|N|-1)\theta_i + \{ \sum_{j \in P_i(\sigma)} \frac{\theta_j}{2} + (|N|-\sigma_i)\frac{\theta_i}{2} \} + (\sigma_i - 1)\theta_i \\ &= -(|N|-\sigma_i)\theta_i + \{ \sum_{j \in P_i(\sigma)} \frac{\theta_j}{2} + (|N|-\sigma_i)\frac{\theta_i}{2} \} \\ &= \sum_{j \in P_i(\sigma)} \frac{\theta_j}{2} - (|N|-\sigma_i)\frac{\theta_i}{2}, \end{aligned}$$

the desired conclusion. □

Remark 3.1 There is an interesting class of problems for which two parallel perspectives can be taken. Let $N \in \mathcal{N}$ be given. A bankruptcy problem consists of a pair (c, E) where $c \in \mathbf{R}^N_+$ is a claims vector and $E \in \mathbf{R}_+$ is an amount to divide. The amount to divide E is not sufficient to honor all claims. Once again, we can take two different approaches. If we take the optimistic approach, then the coalitional members have priority over the non-coalitional members, and the worth of each coalition is defined by setting, for all $S \subseteq N$, $v_O(S) = \min\{\sum_{i \in S} c_i, E\}$. However, if we take the pessimistic approach, then the non-coalitional members have priority over the coalitional members, and the worth of each coalition is defined by setting, for all $S \subseteq N$, $v_P(S) = \max\{E - \sum_{i \in N \setminus S} c_i, 0\}$. Since these two formulations are dual to each other, they give the same allocation when the Shapley value is applied. However, this is not the case for the queueing problems considered here.[3]

3.5 Coincidence of the Shapley Value and the Nucleolus

We apply another well-known solution for cooperative games, the nucleolus (Schmeidler 1969), to pessimistic queueing games, and identify the resulting rule. Surprisingly, we end up with the same rule: The Shapley value and the nucleolus coincide for pessimistic queueing games (Chun and Hokari 2007). First, we introduce an *auxiliary pessimistic queueing game* $\tilde{v}_P$, in which the worth of

[3]See Aumann and Maschler (1985), Driessen (1998), and Thomson (2003) for details.

coalition S is obtained by adding $\sum_{i\in S}(n-1)\theta_i$ to $v_P(S)$, that is, for all $S \subseteq N$, $\tilde{v}_P(S) = v_P(S)+\sum_{i\in S}(n-1)\theta_i$. Obviously, $\tilde{v}_P$ is a game in Γ^N and satisfies the zero-normalized condition, that is, for all $i \in N$, $\tilde{v}_P(\{i\}) = 0$. Moreover, $\tilde{v}_P$ is convex and its nucleolus is well-defined. If we show the coincidence of the Shapley value and the nucleolus for auxiliary pessimistic queueing games, then the coincidence for the pessimistic queueing games follows from the fact that both the Shapley value and the nucleolus satisfy *zero-independence*, requiring that adding a constant to the worth of coalitions containing player i should affect her payoff by the constant.

Before we apply the nucleolus to auxiliary pessimistic queueing games and investigate what recommendation it makes, we show that the worth of a coalition with more than two members can be expressed as a sum of worths of two-person coalitions. It can easily be proven from the facts that (1) for all $i \in N$, $\tilde{v}_P(\{i\}) = 0$ and (2) for all $i,j \in N$, $\tilde{v}_P(\{i,j\}) = \max\{\theta_i, \theta_j\}$.

Lemma 3.1 *For all $\theta \in \mathcal{Q}^N$, its auxiliary pessimistic queueing game $\tilde{v}_P$ satisfies*

(i) for all $i \in N$, $\tilde{v}_P(\{i\}) = 0$,
(ii) for all $S \subseteq N$ with $|S| \geq 2$, $\tilde{v}_P(S) = \sum_{T\subseteq S, |T|=2} \tilde{v}_P(T)$ and $\tilde{v}_P(S) \geq 0$.

Now we present an example showing how the worth of a coalition is calculated.

Example 3.1 Let $N \equiv \{1,2,3,4\}$ and $\theta \in \mathbf{R}^N_+$ with $\theta_1 \geq \theta_2 \geq \theta_3 \geq \theta_4$. Then,

$$\begin{aligned}
\tilde{v}_P(\{1,2,3\}) = 2\theta_1 + \theta_2 &= \tilde{v}_P(\{1,2\}) + \tilde{v}_P(\{1,3\}) + \tilde{v}_P(\{2,3\}),\\
\tilde{v}_P(\{1,2,4\}) = 2\theta_1 + \theta_2 &= \tilde{v}_P(\{1,2\}) + \tilde{v}_P(\{1,4\}) + \tilde{v}_P(\{2,4\}),\\
\tilde{v}_P(\{1,3,4\}) = 2\theta_1 + \theta_3 &= \tilde{v}_P(\{1,3\}) + \tilde{v}_P(\{1,4\}) + \tilde{v}_P(\{3,4\}),\\
\tilde{v}_P(\{2,3,4\}) = 2\theta_2 + \theta_3 &= \tilde{v}_P(\{2,3\}) + \tilde{v}_P(\{2,4\}) + \tilde{v}_P(\{3,4\}),\\
\tilde{v}_P(\{1,2,3,4\}) = 3\theta_1 + 2\theta_2 + \theta_3 &= \tilde{v}_P(\{1,2\}) + \tilde{v}_P(\{1,3\})\\
&\quad +\tilde{v}_P(\{1,4\}) + \tilde{v}_P(\{2,3\}) + \tilde{v}_P(\{2,4\}) + \tilde{v}_P(\{3,4\}).
\end{aligned}$$

Let $\tilde{\Gamma}^N$ be the class of games satisfying the two conditions of Lemma 3.1. That is, $v \in \tilde{\Gamma}^N$ if and only if for all $i \in N$, $v(\{i\}) = 0$, and for all $S \subseteq N$ with $|S| \geq 2$, $v(S) = \sum_{T\subseteq S,|T|=2} v(T)$ and $v(S) \geq 0$. This class includes, in particular, our auxiliary pessimistic queueing games and more. From Deng and Papadimitriou (1994) and van den Nouweland et al. (1996), the coincidence between the Shapley value and the nucleolus can be established.

Here, we present a proof using the Kohlberg's lemma (Kohlberg 1971). First, we show that the Shapley value of the auxiliary pessimistic queueing game can be calculated by using only the worths of the two-person coalitions: It assigns to each agent a half of the sum of her contributions on all two-person coalitions. We note that its computational burden is significantly reduced since we need to know $n(n-1)/2$ numbers instead of $2^n - 1$ numbers.

Lemma 3.2 *For all $v \in \tilde{\Gamma}^N$ and all $i \in N$, $Sh_i(v) = \frac{1}{2} \sum_{S \subseteq N, S \ni i, |S|=2} v(S)$.*

Proof Let $v \in \tilde{\Gamma}^N$, $S \subseteq N$ be such that $|S| \geq 2$, and $i \in S$. Note that

$$\begin{aligned} v(S) - v(S \setminus \{i\}) &= \sum_{T \subseteq S, |T|=2} v(T) - \sum_{R \subseteq S \setminus \{i\}, |R|=2} v(R) \\ &= \sum_{T \subseteq S, T \ni i, |T|=2} v(T). \end{aligned}$$

Thus, for all $i \in N$,

$$\begin{aligned} Sh_i(v) &= \sum_{S \subseteq N, S \ni i} \frac{(|S|-1)!(|N|-|S|)!}{|N|!} \cdot \sum_{T \subseteq S, T \ni i, |T|=2} v(T) \\ &= \sum_{S \subseteq N, S \ni i} \sum_{T \subseteq S, T \ni i, |T|=2} \frac{(|S|-1)!(|N|-|S|)!}{|N|!} \cdot v(T) \\ &= \sum_{T \subseteq N, T \ni i, |T|=2} \sum_{S \subseteq N, S \supseteq T} \frac{(|S|-1)!(|N|-|S|)!}{|N|!} \cdot v(T) \\ &= \sum_{T \subseteq N, T \ni i, |T|=2} \sum_{k=0}^{|N|-2} \frac{(|N|-2)!}{k!(|N|-2-k)!} \cdot \frac{(k+1)!(|N|-k-2)!}{|N|!} \cdot v(T) \\ &= \sum_{T \subseteq N, T \ni i, |T|=2} \sum_{k=0}^{|N|-2} \frac{k+1}{|N| \cdot (|N|-1)} \cdot v(T) \\ &= \sum_{T \subseteq N, T \ni i, |T|=2} \frac{1+2+\cdots+(|N|-1)}{|N| \cdot (|N|-1)} \cdot v(T) \\ &= \sum_{T \subseteq N, T \ni i, |T|=2} \frac{v(T)}{2}, \end{aligned}$$

the desired conclusion. □

Now we show that at the Shapley value allocation, the excess of a coalition equals to the excess of its complementary coalition.

Lemma 3.3 *For all $v \in \tilde{\Gamma}^N$ and all $i \in N$, if $x_i = \frac{1}{2} \sum_{S \subseteq N, S \ni i, |S|=2} v(S)$, then for all $S \subseteq N$,*

$$v(S) - \sum_{i \in S} x_i = v(N \setminus S) - \sum_{i \in N \setminus S} x_i.$$

Proof Let $v \in \tilde{\Gamma}^N$ and $S \subset N$. If $1 < |S| < |N|$, then

$$\begin{aligned} v(S) - \sum_{i \in S} x_i &= \sum_{T \subseteq S, |T|=2} v(T) - \sum_{i \in S} \frac{1}{2} \sum_{T \subseteq N, T \ni i, |T|=2} v(T) \\ &= -\frac{1}{2} \sum_{\{i,j\} \subseteq N, i \in S, j \in N \setminus S} v(\{i,j\}), \end{aligned}$$

and

$$\begin{aligned} v(N \setminus S) - \sum_{j \in N \setminus S} x_j &= \sum_{T \subseteq N \setminus S, |T|=2} v(T) - \sum_{j \in N \setminus S} \frac{1}{2} \sum_{T \subseteq N, T \ni j, |T|=2} v(T) \\ &= -\frac{1}{2} \sum_{\{i,j\} \subseteq N, i \in S, j \in N \setminus S} v(\{i,j\}). \end{aligned}$$

If $S = N \setminus \{j\}$, then

$$\begin{aligned} v(S) - \sum_{i \in S} x_i &= \sum_{T \subseteq S, |T|=2} v(T) - \sum_{i \in S} \frac{1}{2} \sum_{T \subseteq N, T \ni i, |T|=2} v(T) \\ &= -\frac{1}{2} \sum_{\{i,j\} \subseteq N, i \in S} v(\{i,j\}) \\ &= v(\{j\}) - x_j, \end{aligned}$$

the desired conclusion. □

For all $v \colon 2^N \to \mathbf{R}$, all $x \in \mathbf{R}^N$ with $\sum_{i \in N} x_i = v(N)$, and all $\alpha \in \mathbf{R}$, let

$$\mathscr{S}_\alpha(v, x) \equiv \left\{ S \in 2^N \;\middle|\; S \neq \emptyset \text{ and } v(S) - \sum_{i \in S} x_i \geq \alpha \right\}.$$

A collection $\mathscr{B} \subseteq 2^N$ of coalitions is *strictly balanced on N* if there exists a list $(\delta_S)_{S \in \mathscr{B}}$ of positive weights such that for all $i \in N$,

$$\sum_{S \in \mathscr{B}, S \ni i} \delta_S = 1.$$

Lemma 3.4 (Kohlberg 1971) *For all* $v \in \Gamma^N$ *and all* $x \in I(v)$,

$$x = Nu(v) \Leftrightarrow \left[\begin{array}{l}\text{for all } \alpha \in \mathbf{R} \text{ with } \mathscr{S}_\alpha(v,x) \neq \emptyset, \\ \text{there exists } \mathscr{S} \subseteq \big\{\{i\} \mid i \in N \text{ and } v(\{i\}) - x_i = 0\big\} \\ \text{such that } \mathscr{S}_\alpha(v,x) \cup \mathscr{S} \text{ is strictly balanced on } N.\end{array}\right]$$

We are ready to state and prove our coincidence result.

Theorem 3.3 (Chun and Hokari 2007) *For all* $v \in \tilde{\Gamma}^N$, $Sh(v) = Nu(v)$.

Proof By Lemma 3.2, for all $v \in \tilde{\Gamma}^N$ and all $i \in N$,

$$Sh_i(v) = \frac{1}{2} \sum_{S \subseteq N, S \ni i, |S|=2} v(S).$$

Let $\alpha \in \mathbf{R}$ be such that $\mathscr{S}_\alpha(v, Sh(v)) \neq \emptyset$. Let $S \in \mathscr{S}_\alpha(v, Sh(v))$. Since by Lemma 3.3,

$$v(N \setminus S) - \sum_{i \in N \setminus S} Sh_i(v) = v(S) - \sum_{i \in S} Sh_i(v),$$

$N \setminus S \in \mathscr{S}_\alpha(v, Sh(v))$. Thus, $\mathscr{S}_\alpha(v, Sh(v))$ is strictly balanced on N. The desired conclusion follows from Lemma 3.4. □

We can obtain a similar conclusion for the optimistic queueing game. Since for all $S \subseteq N$, $v_O(S) \leq 0$, this game does not satisfy the conditions of Lemma 3.1 and moreover, its core is empty. However, it is not difficult to show that this game satisfies other two conditions, that is,

(i) for all $i \in N$, $v_O(\{i\}) = 0$,
(ii) for all $S \subseteq N$ with $|S| \geq 2$, $v_O(S) = \sum_{T \subseteq S, |T|=2} v_O(T)$.

As shown in Kar et al. (2009), these two conditions are sufficient to guarantee the coincidence of the Shapley value and the prenucleolus.

3.6 Other Rules Applied to Queueing Games

We discuss the relation between the minimal and the maximal transfer rules and other rules discussed in the literature, the core, the τ-value, and the serial cost sharing rule.

3.6.1 The Core

For a convex game, both the Shapley value and the nucleolus select an allocation in the core. It is natural to ask about the structure of the core for auxiliary pessimistic queueing games. In particular, one might conjecture that the coincidence between the two solutions comes from the fact that the core is a singleton. As shown in Fig. 3.1 for a three-agent problem with $N \equiv \{1,2,3\}$ and $\theta_1 \geq \theta_2 \geq \theta_3$, this is not the case. Its core is pretty large. However, it has a rather symmetric structure. This is the central reason why we obtain the coincidence of the two solutions.

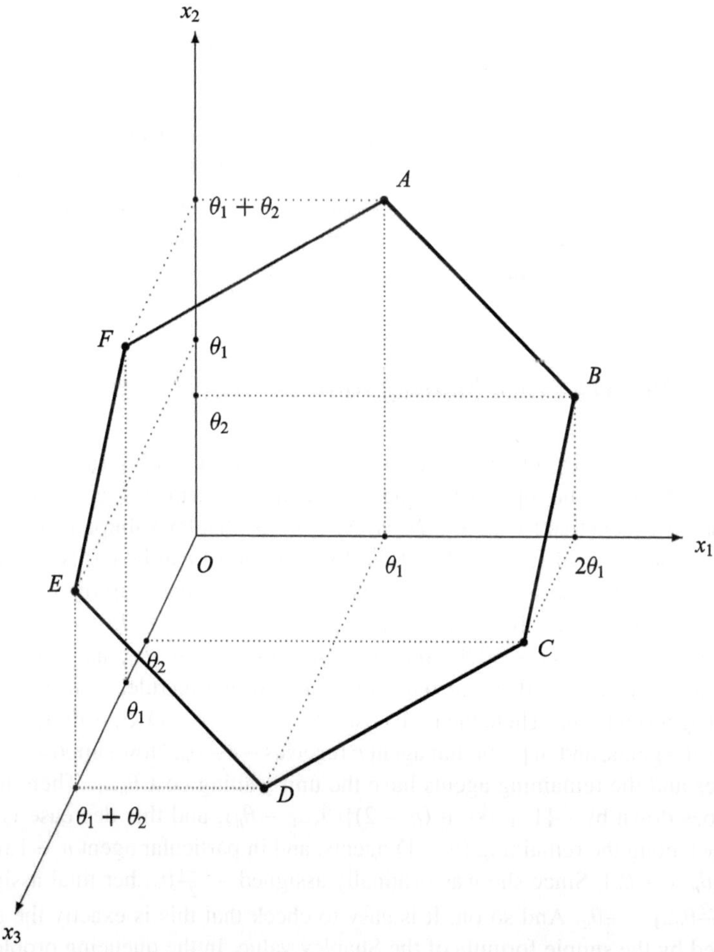

Fig. 3.1 The core of an auxiliary pessimistic queueing game may not be completely symmetric, but it is sufficiently symmetric to guarantee $Sh(\widetilde{v}_P) = Nu(\widetilde{v}_P)$. In the figure, the core is the interior (including the boundary) of *ABCDEF*

3.6.2 The τ-value

For all $v \in \Gamma^N$ and all $i \in N$, let $M_i(v) \equiv v(N) - v(N \setminus \{i\})$ and $m_i(v) \equiv v(\{i\})$. Then, the τ-value (Tijs 1987) selects the maximal feasible allocation on the line connecting $M(v) \equiv (M_i(v))_{i \in N}$ and $m(v) \equiv (m_i(v))_{i \in N}$.

τ-value, τ: For each convex game v,

$$\tau(v) \equiv \lambda M(v) + (1 - \lambda) m(v),$$

where $\lambda \in [0, 1]$ is chosen so as to satisfy

$$\sum_{j \in N} \left[\lambda\big(v(N) - v(N \setminus \{j\})\big) + (1 - \lambda) v(\{j\})\right] = v(N).$$

The auxiliary pessimistic queueing game $\tilde{v}_P$ is convex and $m(\tilde{v}_P) = 0$. Moreover, it is easy to see that for all $j \in N$, $\tilde{v}_P(N) - \tilde{v}_P(N \setminus \{j\}) = \sum_{S \ni j, |S|=2} \tilde{v}_P(S)$ and that $\lambda = \frac{1}{2}$. By using the fact that the τ-value satisfies *zero-independence*, we can establish the coincidence of the τ-value, the Shapley value, and the nucleolus for pessimistic queueing games.

3.6.3 The Serial Cost Sharing Rule

The minimal transfer rule coincides with the serial cost sharing rule for queueing problems.[4] Since the optimistic queueing game v_O satisfies all conditions of Lemma 3.1 except that for all $S \subseteq N$, $v_O(S) \geq 0$, its Shapley value can be expressed in the simple form of Lemma 3.2. Therefore, the proof can be easily obtained by checking the simple formula for the Shapley value. To further simplify the argument, let $N \in \mathcal{N}$ and $\theta \in \mathcal{Q}^N$ be such that $N = \{1, \ldots, n\}$ and $\theta_1 \geq \theta_2 \geq \cdots \geq \theta_n$. From Lemma 3.2, $Sh_n(v_O) = -\frac{n-1}{2}\theta_n$, $Sh_{n-1}(v_O) = -\frac{n-2}{2}\theta_{n-1} - \frac{1}{2}\theta_n$, and so on.

To calculate the payoff assigned by the serial cost sharing rule, we need to assume that all agents have θ_n. Then, the total cost $-\{1 + \cdots + (n-1)\}\theta_n$ is divided equally among all agents, and in particular agent n receives $-\frac{n-1}{2}\theta_n$. Now suppose that agent n leaves and the remaining agents have the unit waiting cost θ_{n-1}. Then, the total cost goes down by $-\{1 + \cdots + (n-2)\}(\theta_{n-1} - \theta_n)$, and this decrease is shared equally among the remaining $(n-1)$ agents, and in particular agent $n-1$ receives $-\frac{n-2}{2}(\theta_{n-1} - \theta_n)$. Since she was originally assigned $-\frac{n-1}{2}\theta_n$, her final assignment is $-\frac{n-2}{2}\theta_{n-1} - \frac{1}{2}\theta_n$. And so on. It is easy to check that this is exactly the amount assigned by the simple formula of the Shapley value. In the queueing problem, the serial cost sharing and the minimal transfer rules make the same recommendation.

[4] Moulin (2007) makes the same observation for the scheduling problem.

On the other hand, the maximal transfer rule coincides with the decreasing serial cost sharing rule (de Frutos 1998) for queueing problems. Once again, by Lemma 3.2, the Shapley value can be expressed in the simple form and the proof can be easily obtained by checking the simple formula for the Shapley value. To simplify the argument, let $N \in \mathcal{N}$ and $\theta \in \mathcal{Q}^N$ be such that $N = \{1, \ldots, n\}$ and $\theta_1 \geq \theta_2 \geq \cdots \geq \theta_n$. From Lemma 3.2, $Sh_1(v_P) = -\frac{n-1}{2}\theta_1$, $Sh_2(v_P) = -\frac{1}{2}\theta_1 - \frac{n-2}{2}\theta_2$, and so on.

To calculate the payoff assigned by the decreasing serial cost sharing rule, we need to assume that all agents have θ_1. Then, the total cost $-\{1 + \cdots + (n-1)\}\theta_1$ is divided equally among all agents, and in particular agent 1 receives $-\frac{n-1}{2}\theta_1$. Now suppose that agent 1 leaves and the remaining agents have the unit waiting cost θ_2. Then, the total cost goes up by $\{1+\cdots+(n-2)\}(\theta_1-\theta_2)$, and this increase is shared equally among the remaining $(n-1)$ agents, and in particular agent $n-1$ receives $\frac{n-2}{2}(\theta_1 - \theta_2)$. Since she was originally assigned $-\frac{n-1}{2}\theta_1$, her final assignment is $-\frac{1}{2}\theta_1 - \frac{n-2}{2}\theta_2$. And so on. It is easy to check that this is exactly the amount assigned by the simple formula of the Shapley value. In the queueing problem, the decreasing serial cost sharing and the minimal transfer rules make the same recommendation.

3.7 Concluding Remarks

In this chapter, we pointed out the importance of the way in which the worth of a coalition is defined when problems are mapped into cooperative games. In particular, depending upon who will be served first, two different definitions of the worth of a coalition can be obtained, and these definitions lead to very different rules. For other classes of problems such as bankruptcy, the two perspectives give the same recommendation. It would be interesting to develop a general theory to explain the relation.

References

Aumann, R., & Maschler, M. (1985). Game theoretic analysis of a bankruptcy problem from the Talmud. *Journal of Economic Theory, 36*, 195–213.

Chun, Y. (2006). A pessimistic approach to the queueing problem. *Mathematical Social Sciences, 51*, 171–181.

Chun, Y., & Hokari, T. (2007). On the coincidence of the Shapley value and the nucleolus in queueing problems. *Seoul Journal of Economics, 20*(2), 223–237.

de Frutos, M. A. (1998). Decreasing serial cost sharing under economies of scale. *Journal of Economic Theory, 79*, 245–275.

Deng, X., & Papadimitriou, C. H. (1994). On the complexity of cooperative solution concepts. *Mathematics of Operations Research, 19*, 257–266.

Driessen, T. (1998). The greedy bankruptcy game: An alternative game theoretic analysis of a bankruptcy problem. In L. A. Petrosjan & V. V. Mazalov (Eds.), *Game theory and applications IV*. New York: Nova Science Publishers Inc.

Kar, A., Mitra. M., & Mutuswami, S. (2009). On the coincidence of the prenucleolus and the Shapley value. *Mathematical Social Sciences, 57*, 16–25.
Kohlberg, E. (1971). On the nucleolus of a characteristic function game. *SIAM Journal on Applied Mathematics, 20*, 62–66.
Maniquet, F. (2003). A characterization of the Shapley value in queueing problems. *Journal of Economic Theory, 109*, 90–103.
Moulin, H. (2007). On scheduling fees to prevent merging, splitting, and transferring of jobs. *Mathematics of Operations Research, 32*, 266–283.
Moulin, H., & Shenker, S. (1992). Serial cost sharing. *Econometrica, 50*(5), 1009–1039.
Schmeidler, D. (1969). The nucleolus of a characteristic function game. *SIAM Journal on Applied Mathematics, 17*, 1163–1170.
Shapley, L. S. (1953). A value for n-person games. In H. W. Kuhn & A. W. Tucker (Eds.), *Contributions to the theory of games II*. Annals of mathematics studies (Vol. 28, pp. 307–317). Princeton: Princeton University Press.
Thomson, W. (2003). How to divide when there isn't enough: From the Talmud to game theory. University of Rochester, Book Manuscript.
Tijs, S. H. (1987). An axiomatization of the τ-value. *Mathematical Social Sciences, 13*, 177–181.
van den Nouweland, A., Borm, P., van Golstein Brouwers, W., Groot Bruinderink, R., & Tijs, S. H. (1996). A game theoretic approach to problems in telecommunication. *Management Science, 42*, 294–303.

Chapter 4
Independence, Monotonicity, and Balanced Consistency

4.1 Introduction

In this chapter, we present characterizations of the minimal and the maximal transfer rules on the basis of various axioms specifying how a rule should respond to changes in the waiting cost or population. We begin with two independence axioms which require that a change in an agent's waiting cost should not affect the agents following her or preceding her. More precisely, suppose that an agent's waiting cost changes. One could take two different perspectives with regard to how the allocation should be affected by this change: (1) an increase in an agent's waiting cost would affect her predecessors, but not her followers, or (2) a decrease in an agent's waiting cost would affect her followers, but not her predecessors. *Independence of preceding costs* (Maniquet 2003) requires that an increase in an agent's waiting cost should not affect the agents following her. On the other hand, *independence of following costs* (Chun 2006) requires that a decrease in an agent's waiting cost should not affect the agents preceding her. The minimal transfer rule satisfies *independence of preceding costs,* but the maximal transfer rule satisfies *independence of following costs.* Moreover, together with basic axioms, the minimal transfer rule is the only rule satisfying *independence of preceding costs,* whereas the maximal transfer rule is the only rule satisfying *independence of following costs.*

Next are two solidarity and two equal responsibility requirements. Solidarity axioms require that all agents should gain together or lose together as a consequence of changes in their external environment. More precisely, if there is an increase in the waiting cost of an agent, then one could take two different perspectives with regard to how the allocation should be affected by this increase: (1) one may feel that she deserves greater compensation for her waiting, which will affect other agents in a negative direction, or (2) one may feel that she should be required to pay more for having the service, which will affect other agents in a positive direction. *Negative cost monotonicity* (Maniquet 2003) requires that an increase in an agent's waiting cost should cause all other agents to weakly lose. On the other hand, *positive cost*

© Springer International Publishing Switzerland 2016

Y. Chun, *Fair Queueing*, Studies in Choice and Welfare,
DOI 10.1007/978-3-319-33771-5_4

monotonicity (Chun 2006) requires that an increase in an agent's waiting cost should cause all other agents to weakly gain.

Equal responsibility axioms are concerned with changes in the population. If some agent in the queue leaves, then under *queue-efficiency,* the queue is affected minimally, that is, her precedents remain at the same position, but her followers move forward by one position. However, the monetary compensations may need to be adjusted. *Last-agent equal responsibility* (Maniquet 2003) requires that upon the departure of the agent served last, all other agents should remain at the same position and their transfers should be affected by the same amount. On the other hand, *first-agent equal responsibility* (Chun 2006) requires that upon the departure of the agent served first, all other agents should move forward by one position and their transfers should be affected by the same amount. The minimal transfer rule satisfies *negative cost monotonicity* and *last-agent equal responsibility,* but the maximal transfer rule satisfies *positive cost monotonicity* and *first-agent equal responsibility*. Moreover, together with basic axioms, each rule is the only rule satisfying the axioms.

We also investigate the implications of *balanced consistency* and *balanced cost reduction* (van den Brink and Chun 2012). *Balanced consistency* requires that the effect on the payoff from the departure of one agent to another agent should be equal between any two agents. On the other hand, *balanced cost reduction* requires that if one agent leaves a problem, then the total payoffs of the remaining agents should be affected by the amount previously assigned to the leaving agent. We show that the minimal transfer rule is the only rule satisfying *efficiency* and *Pareto indifference* together with either one of the two axioms, *balanced consistency* and *balanced cost reduction.* Upon the departure of an agent, if we assume that all of his predecessors are moving back by one position to keep the same completion time, an alternative axiom of *balanced consistency under constant completion time* can be formulated. The maximal transfer rule is the only rule satisfying *efficiency, Pareto indifference,* and *balanced consistency under constant completion time.*

This chapter is organized as follows.[1] In Sect. 4.2, we present characterizations of the minimal and the maximal transfer rules on the basis of *independence* and in Sect. 4.3 on the basis of *monotonicity* and *equal responsibility*. In Sect. 4.4, the two rules are characterized by *balanced consistency* or *balanced cost reduction.* Concluding remarks follow in Sect. 4.5.

4.2 Independence

We begin with two independence axioms which require that a change in an agent's waiting cost should not affect the agents following her or preceding her. More precisely, suppose that an agent's waiting cost changes. One could take two different

[1]The results of this chapter are collected from Maniquet (2003), Chun (2006), and van den Brink and Chun (2012). Excerpts from Chun (2006) are reprinted with kind permission of Elsevier.

perspectives with regard to how the allocation should be affected by this change: (1) an increase in an agent's waiting cost would affect her predecessors, but not her followers, or (2) a decrease in an agent's waiting cost would affect her followers, but not her predecessors. *Independence of preceding costs* requires that an increase in an agent's waiting cost should not affect the agents following her. On the other hand, *independence of following costs* requires that a decrease in an agent's waiting cost should not affect the agents preceding her.

Independence of preceding costs: For all $N \in \mathcal{N}$, all $\theta, \theta' \in \mathcal{Q}^N$, all $(\sigma, t) \in \varphi(\theta)$, all $(\sigma', t') \in \varphi(\theta')$, and all $k \in N$, if for all $i \in N \backslash \{k\}$, $\theta_i = \theta'_i$ and $\theta_k < \theta'_k$, then for all $j \in N$ such that $\sigma_j > \sigma_k$, $u_j(\sigma_j, t_j; \theta_j) = u_j(\sigma'_j, t'_j; \theta'_j)$.

Independence of following costs: For all $N \in \mathcal{N}$, all $\theta, \theta' \in \mathcal{Q}^N$, all $(\sigma, t) \in \varphi(\theta)$, all $(\sigma', t') \in \varphi(\theta')$, and all $k \in N$, if for all $i \in N \backslash \{k\}$, $\theta_i = \theta'_i$ and $\theta_k > \theta'_k$, then for all $j \in N$ such that $\sigma_j < \sigma_k$, $u_j(\sigma_j, t_j; \theta_j) = u_j(\sigma'_j, t'_j; \theta'_j)$.

We are ready to state and prove our characterization results on the basis of independence requirements.

Theorem 4.1

(1) (Maniquet 2003) The minimal transfer rule is the only rule satisfying efficiency, Pareto indifference, equal treatment of equals, and independence of preceding costs.[2]

(2) (Chun 2006) The maximal transfer rule is the only rule satisfying efficiency, Pareto indifference, equal treatment of equals, and independence of following costs.

Proof

(1) It is clear that the minimal transfer rule φ^M satisfies *efficiency, Pareto indifference,* and *equal treatment of equals.* The fact that φ^M satisfies *independence of preceding costs* comes from the definition of t_i^M, which does not depend on the waiting costs of the precedents.

Conversely, let φ be a rule satisfying the four axioms. Let $N \in \mathcal{N}$ be such that $N = \{1, \ldots, n\}$, $\theta \in \mathcal{Q}^N$, and $(\sigma, t) \in \varphi(\theta)$. We may assume without loss of generality that $\theta_1 \geq \theta_2 \geq \cdots \geq \theta_n$. Let $\theta^1 \in \mathcal{Q}^N$ be such that for all $i \in N$, $\theta_i^1 = \theta_n$, and $(\sigma^1, t^1) \in \varphi(\theta^1)$. By *efficiency* and *equal treatment of equals,* $u_n(\sigma_n^1, t_n^1; \theta_n) = -\frac{|N|-1}{2}\theta_n$. By *Pareto indifference,* we may assume that $\sigma_n^1 = n$ and $t_n^1 = -\frac{|N|-1}{2}\theta_n$. For $j = 2, \ldots, n$, let $\theta^j \in \mathcal{Q}^N$ be defined by setting for all $i \in N$, $\theta_i^j = \theta_j$ if $i \leq j$, and $\theta_i^j = \theta_i$ otherwise. Let $(\sigma^j, t^j) \in \varphi(\theta^j)$. By *independence of preceding costs,* $u_i(\sigma_i^j, t_i^j; \theta_i^j) = u_i(\sigma_i^{j-1}, t_i^{j-1}; \theta_i^{j-1})$ for all

[2] Instead of *Pareto indifference,* Maniquet (2003) imposes *anonymity,* which requires that relabeling of agents should not affect the allocation chosen by a rule. Since the same result can be obtained by imposing *Pareto indifference* (with the same proof), we impose *Pareto indifference* here.

$i > j$. By *efficiency* and *equal treatment of equals,* for all $i \leq j$, we have

$$u_i(\sigma^j_i, t^j_i; \theta^j_i) = -(j-1)\frac{\theta_j}{2} - \frac{T^j}{j},$$

where $T^j = \sum_{i>j} t^j_i$. By *Pareto indifference,* we may assume that $\sigma^j_j = j$. Applying this formula to $j = n-1, \ldots, 1$, we obtain

$$u_{n-1}(\sigma^1_{n-1}, t^1_{n-1}; \theta_{n-1}) = -\left[(n-2)\frac{\theta_{n-1}}{2} - \frac{\theta_n}{2}\right], \quad t^1_{n-1} = (n-2)\frac{\theta_{n-1}}{2} - \frac{\theta_n}{2},$$

$$u_{n-2}(\sigma^1_{n-2}, t^1_{n-2}; \theta_{n-2}) = -\left[(n-3)\frac{\theta_{n-2}}{2} - \frac{\theta_{n-1}+\theta_n}{2}\right],$$

$$t^1_{n-2} = (n-3)\frac{\theta_{n-2}}{2} - \frac{\theta_{n-1}+\theta_n}{2},$$

$$\vdots$$

$$u_i(\sigma^1_i, t^1_i; \theta_i) = -\left[(i-1)\frac{\theta_i}{2} - \sum_{j>i}\frac{\theta_j}{2}\right], \quad t^1_i = (i-1)\frac{\theta_i}{2} - \sum_{j>i}\frac{\theta_j}{2},$$

the desired conclusion.

(2) It is clear that the maximal transfer rule φ^C satisfies *efficiency, Pareto indifference,* and *equal treatment of equals.* The fact that φ^C satisfies *independence of following costs* comes from the definition of t^C_i, which does not depend on the waiting costs of the followers.

Conversely, let φ be a rule satisfying the four axioms. Let $N \in \mathcal{N}$ be such that $N = \{1, \ldots, n\}$, $\theta \in \mathcal{Q}^N$, and $(\sigma, t) \in \varphi(\theta)$. We may assume without loss of generality that $\theta_1 \geq \theta_2 \geq \cdots \geq \theta_n$. Let $\theta^1 \in \mathcal{Q}^N$ be such that for all $i \in N$, $\theta^1_i = \theta_1$, and $(\sigma^1, t^1) \in \varphi(\theta^1)$. By *efficiency* and *equal treatment of equals,* $u_1(\sigma^1_1, t^1_1; \theta_1) = -\frac{|N|-1}{2}\theta_1$. By *Pareto indifference,* we may assume that $\sigma^1_1 = 1$ and $t^1_1 = -\frac{|N|-1}{2}\theta_1$. For $j = 2, \ldots, n$, let $\theta^j \in \mathcal{Q}^N$ be defined by setting for all $i \in N$, $\theta^j_i = \theta_j$ if $i \geq j$, and $\theta^j_i = \theta_i$ otherwise. Let $(\sigma^j, t^j) \in \varphi(\theta^j)$. By *independence of following costs,* $u_i(\sigma^j_i, t^j_i; \theta^j_i) = u_i(\sigma^{j-1}_i, t^{j-1}_i; \theta^{j-1}_i)$ for all $i < j$. By *efficiency* and *equal treatment of equals,* for all $i \geq j$, we have

$$u_i(\sigma^j_i, t^j_i; \theta^j_i) = -\frac{|N|+j-2}{2}\theta_j - \frac{T^j}{|N|-j+1},$$

where $T^j = \sum_{i<j} t^j_i$. By *Pareto indifference,* we may assume that $\sigma^j_j = j$. Applying this formula to $j = 2, \ldots, n$, we obtain

$$u_2(\sigma_2^1, t_2^1; \theta_2) = -\frac{|N|}{2}\theta_2 + \frac{1}{2}\theta_1, \quad t_2^1 = -\frac{|N|-2}{2}\theta_2 + \frac{1}{2}\theta_1,$$

$$u_3(\sigma_3^1, t_3^1; \theta_3) = -\frac{|N|+1}{2}\theta_3 + \frac{1}{2}(\theta_1 + \theta_2), \quad t_3^1 = -\frac{|N|-3}{2}\theta_3 + \frac{1}{2}(\theta_1 + \theta_2),$$

$$\vdots$$

$$u_i(\sigma_i^1, t_i^1; \theta_i) = -\frac{|N|+i-2}{2}\theta_i + \frac{1}{2}\sum_{j<i}\theta_j, \quad t_i^1 = -\frac{|N|-i}{2}\theta_i + \frac{1}{2}\sum_{j<i}\theta_j,$$

the desired conclusion. □

4.3 Monotonicity and Equal Responsibility

Now we introduce two solidarity requirements that all agents should gain together or lose together as a consequence of changes in their external environment. Axioms in this spirit have been discussed in the context of a wide range of problems, under various names: *resource monotonicity* requires that a change in the resources should affect all agents in the same direction (Chun and Thomson 1988; Moulin and Thomson 1988; Roemer 1986, and others) and *welfare domination under preference replacement* requires that a change in an agent's preference should affect all the others in the same direction (Moulin 1987; Thomson 1999, and others).

Returning to our current model, suppose that the waiting cost of one agent increases. One could take two different perspectives with regard to how the allocation should be affected by this change: (1) one may feel that she deserves greater compensation for her waiting, which will affect other agents in a negative direction, or (2) one may feel that she should be required to pay more for having the service, which will affect other agents in a positive direction.

Negative cost monotonicity requires that an increase in an agent's waiting cost should cause all other agents to weakly lose. On the other hand, *positive cost monotonicity* requires that an increase in an agent's waiting cost should cause all other agents to weakly gain.

Negative cost monotonicity: For all $N \in \mathcal{N}$, all $\theta, \theta' \in \mathcal{Q}^N$, all $(\sigma, t) \in \varphi(\theta)$, all $(\sigma', t') \in \varphi(\theta')$, and all $k \in N$, if for all $i \in N\backslash\{k\}$, $\theta_i = \theta'_i$ and $\theta_k < \theta'_k$, then for all $i \in N\backslash\{k\}$, $u_i(\sigma_i, t_i; \theta_i) \geq u_i(\sigma'_i, t'_i; \theta'_i)$.

Positive cost monotonicity: For all $N \in \mathcal{N}$, all $\theta, \theta' \in \mathcal{Q}^N$, all $(\sigma, t) \in \varphi(\theta)$, all $(\sigma', t') \in \varphi(\theta')$, and all $k \in N$, if for all $i \in N\backslash\{k\}$, $\theta_i = \theta'_i$ and $\theta_k < \theta'_k$, then for all $i \in N\backslash\{k\}$, $u_i(\sigma_i, t_i; \theta_i) \leq u_i(\sigma'_i, t'_i; \theta'_i)$.

Next two axioms are concerned with changes in the population. If some agent in the queue leaves, then under *queue-efficiency*, the queue is affected minimally, that is, her precedents remain at the same position, but her followers move forward by one position. However, the monetary compensations may need to be adjusted. *Last-agent equal responsibility* requires that upon the departure of the agent served last, all other agents should remain at the same position and their transfers should be affected by the same amount. On the other hand, *first-agent equal responsibility* requires that upon the departure of the agent served first, all other agents should move forward by one position and their transfers should be affected by the same amount. For all $N \in \mathcal{N}$, all $\theta \in \mathcal{Q}^N$, and all $k \in N$, let $\theta_{N\setminus\{k\}} = (\theta_i)_{i \in N\setminus\{k\}}$. Note that $\theta_{N\setminus\{k\}} \in \mathcal{Q}^{N\setminus\{k\}}$.

Last-agent equal responsibility: For all $N \in \mathcal{N}$, all $\theta \in \mathcal{Q}^N$, and all $(\sigma, t) \in \varphi(\theta)$, if agent $k \in N$ is such that $\sigma_k = |N|$, then there exists $(\sigma', t') \in \varphi(\theta_{N\setminus\{k\}})$ such that for all $i \in N\setminus\{k\}$, $\sigma'_i = \sigma_i$ and $t'_i = t_i + \frac{t_k}{|N|-1}$.

First-agent equal responsibility: For all $N \in \mathcal{N}$, all $\theta \in \mathcal{Q}^N$, and all $(\sigma, t) \in \varphi(\theta)$, if agent $k \in N$ is such that $\sigma_k = 1$, then there exists $(\sigma', t') \in \varphi(\theta_{N\setminus\{k\}})$ such that for all $i \in N\setminus\{k\}$, $\sigma'_i = \sigma_i - 1$ and $t'_i = t_i + \frac{t_k}{|N|-1}$.

Our second characterizations of the two rules are based on *monotonicity* and *equal responsibility* axioms.

Theorem 4.2

(1) *(Maniquet 2003) The minimal transfer rule is the only rule satisfying Pareto indifference, the identical preferences lower bound, negative cost monotonicity, and last-agent equal responsibility.*

(2) *(Chun 2006) The maximal transfer rule is the only rule satisfying Pareto indifference, the identical preferences lower bound, positive cost monotonicity, and first-agent equal responsibility.*

Proof

(1) To simplify notation, we do not attach the superscript M to σ and t. First, we show that φ^M satisfies the *identical preferences lower bound.* Let $N \in \mathcal{N}$, $\theta \in \mathcal{Q}^N$ and $(\sigma, t) \in \varphi^M(\theta)$. Since φ^M satisfies *efficiency,* for all $i \in N$, $-\sum_{j \in F_i(\sigma)} \frac{\theta_j}{2} \geq -(|N| - \sigma_i)\frac{\theta_i}{2}$, which implies $u_i(\sigma_i, t_i; \theta_i) \geq -(|N| - 1)\frac{\theta_i}{2}$.

Next we show that φ^M satisfies *negative cost monotonicity.* Let $N \in \mathcal{N}$, θ, $\theta' \in \mathcal{Q}^N$, $(\sigma, t) \in \varphi^M(\theta)$, $(\sigma', t') \in \varphi^M(\theta')$, and $k \in N$ be such that for all $i \in N\setminus\{k\}$, $\theta_i = \theta'_i$ and $\theta_k < \theta'_k$. By *efficiency,* $\sigma_k \geq \sigma'_k$ and for all $i \in N\setminus\{k\}$, $\sigma_i > \sigma_k$ implies $\sigma'_i > \sigma'_k$. We partition $N\setminus\{k\}$ into N_1, N_2, N_3 defined by

$$N_1 = \{i \in N | \sigma_k < \sigma_i\},$$
$$N_2 = \{i \in N | \sigma_i < \sigma_k, \sigma'_k < \sigma'_i\}, \text{ and}$$
$$N_3 = \{i \in N | \sigma'_i < \sigma'_k\}.$$

We need to show that for all $i \in N\setminus\{k\}$, $u_i(\sigma_i, t_i; \theta_i) \geq u_i(\sigma'_i, t'_i; \theta'_i)$.

(i) If $i \in N_1$, the desired conclusion comes from the fact that φ^M satisfies *independence of preceding costs.*

(ii) If $i \in N_2$, then by *efficiency,* $\theta_i \geq \theta_k$. Note that for all $j \in F_i(\sigma)\backslash F_i(\sigma')$ or $j \in F_i(\sigma')\backslash F_i(\sigma)$ such that $j \neq k$, $\theta_j = \theta_i$. Therefore,

$$\begin{aligned} u_i(\sigma_i', t_i'; \theta_i) - u_i(\sigma_i, t_i; \theta_i) &= \left[-(\sigma_i' - 1)\frac{\theta_i}{2} - \sum_{j \in F_i(\sigma')} \frac{\theta_j}{2} \right] \\ &\quad - \left[-(\sigma_i - 1)\frac{\theta_i}{2} - \sum_{j \in F_i(\sigma)} \frac{\theta_j}{2} \right] \\ &= -\frac{\theta_i}{2} + \frac{\theta_k}{2} \\ &\leq 0, \end{aligned}$$

the desired conclusion.

(iii) If $i \in N_3$, then by *efficiency,* $\theta_i \geq \theta_k'$. Therefore, $u_i(\sigma_i', t_i'; \theta_i) - u_i(\sigma_i, t_i; \theta_i) = -\frac{\theta_k'}{2} + \frac{\theta_k}{2} < 0$, the desired conclusion.

Now we show that φ^M satisfies *last-agent equal responsibility.* Let $N \in \mathcal{N}$, $\theta \in \mathcal{Q}^N$, and $(\sigma, t) \in \varphi^M(\theta)$. Suppose, without loss of generality, that $\sigma_n = n$. Then, $u_n(\sigma_n, t_n; \theta_n) = -(|N| - 1)\frac{\theta_n}{2}$. Consider $\theta_{N\backslash\{n\}} \in \mathcal{Q}^{N\backslash\{n\}}$. By *efficiency,* for all $i \in N\backslash\{n\}$, there is $(\sigma', t') \in \varphi^M(\theta_{N\backslash\{n\}})$ such that $\sigma_i' = \sigma_i$. Moreover, $t_i' - t_i = \frac{\theta_n}{2}$, the desired conclusion.

Conversely, let φ be a rule satisfying the four axioms. Let $N \in \mathcal{N}$, $\theta \in \mathcal{Q}^N$ and $(\sigma, t) \in \varphi(\theta)$.

Step M-1: For $k \in N$ such that for all $i \in N$, $\theta_k \leq \theta_i$, $u_k(\sigma_k, t_k; \theta_k) = -\frac{|N|-1}{2}\theta_k$. By the *identical preferences lower bound,* $u_k(\sigma_k, t_k; \theta_k) \geq -\frac{|N|-1}{2}\theta_k$. Suppose, by way of contradiction, that the claim does not hold, that is, $u_k(\sigma_k, t_k; \theta_k) > -\frac{|N|-1}{2}\theta_k$. Let $\theta' \in \mathcal{Q}^N$ be such that for all $i \in N$, $\theta_i' = \theta_k$, and $(\sigma', t') \in \varphi(\theta')$. By repeated application of *negative cost monotonicity,* $u_k(\sigma_k', t_k'; \theta_k) > -\frac{|N|-1}{2}\theta_k$. By the *identical preferences lower bound,* for all $i \in N\backslash\{k\}$, $u_i(\sigma_i', t_i'; \theta_i') \geq -\frac{|N|-1}{2}\theta_k$. Altogether,

$$\sum_{i \in N} u_i(\sigma_i', t_i'; \theta_i') > -\frac{|N|(|N| - 1)}{2}\theta_k.$$

On the other hand, *feasibility* of a rule requires that

$$\sum_{i \in N} t_i = \sum_{i \in N} u_i(\sigma_i', t_i'; \theta_i') + \{1 + \cdots + (|N| - 1)\}\theta_k \leq 0,$$

or equivalently,

$$\sum_{i\in N} u_i(\sigma'_i, t'_i; \theta'_i) \leq -\frac{|N|(|N|-1)}{2}\theta_k,$$

a contradiction.

Step M-2: If there are $k, \ell \in N$ such that for all $i \in N\backslash\{k\}$, $\theta_i \geq \theta_k$ and $\sigma_\ell = |N|$, then $\theta_\ell = \theta_k$. From Step M-1, $u_k(\sigma_k, t_k; \theta_k) = -\frac{|N|-1}{2}\theta_k$. Suppose, by way of contradiction, that $\theta_\ell > \theta_k$. By *last-agent equal responsibility,* there is $(\sigma', t') \in \varphi(\theta_{N\backslash\{\ell\}})$ such that $u_k(\sigma'_k, t'_k; \theta_k) = -\frac{|N|-1}{2}\theta_k + \frac{t_\ell}{|N|-1}$. Note that for all $i \in N\backslash\{\ell\}$, $\theta_i \geq \theta_k$. From Step M-1, $u_k(\sigma'_k, t'_k; \theta_k) = -\frac{|N|-2}{2}\theta_k$, which implies $t_\ell = \frac{|N|-1}{2}\theta_k$. Taken together, $u_\ell(\sigma_\ell, t_\ell; \theta_\ell) = -(|N|-1)\theta_\ell + \frac{|N|-1}{2}\theta_k < -\frac{|N|-1}{2}\theta_\ell$, in violation of the *identical preferences lower bound.* Therefore, if $\sigma_\ell = |N|$, then $\theta_\ell = \theta_k$ and $t_\ell = \frac{|N|-1}{2}\theta_\ell$.

Step M-3: Now we show that all agents should end up with the utilities assigned by the minimal transfer rule, beginning with the agent in the last position in the queue. Let $\ell \in N$ be such that $\sigma_\ell = |N|$. From Step M-2, $t_\ell = \frac{|N|-1}{2}\theta_\ell$ and for all $i \in N\backslash\{k\}$, $\theta_i \geq \theta_\ell$. Let $L = N\backslash\{\ell\}$ and $k \in L$ be such that for all $i \in L\backslash\{k\}$, $\theta_i \geq \theta_k$. By *last-agent equal responsibility,* there is $(\sigma', t') \in \varphi(\theta_L)$ such that $\sigma'_k = \sigma_k$. From Step M-1 applied to L and k, $u_k(\sigma'_k, t'_k; \theta_k) = \frac{|L|-1}{2}\theta_k$. Moreover, by *last-agent equal responsibility,*

$$u_k(\sigma'_k, t'_k; \theta_k) = u_k(\sigma_k, t_k; \theta_k) + \frac{\theta_\ell}{2},$$

so that

$$u_k(\sigma_k, t_k; \theta_k) = -(\sigma_k - 1)\frac{\theta_k}{2} - \frac{\theta_\ell}{2}.$$

By repeating the argument, we obtain that for all $i \in N$,

$$u_i(\sigma_i, t_i; \theta_i) = -(\sigma_i - 1)\frac{\theta_i}{2} - \sum_{j\in F_i(\sigma)} \frac{\theta_j}{2},$$

which implies that $t_i = (\sigma_i - 1)\frac{\theta_i}{2} - \sum_{j\in F_i(\sigma)} \frac{\theta_j}{2}$.

Step M-4: By *Pareto indifference,* we obtain the desired conclusion.

(2) To simplify notation, we do not attach the superscript C to σ and t. First, we show that φ^C satisfies the *identical preferences lower bound.* Let $N \in \mathcal{N}$, $\theta \in \mathcal{Q}^N$, and $(\sigma, t) \in \varphi^C(\theta)$. Since φ^C satisfies *efficiency,* for all $i \in N$, $\sum_{j\in P_i(\sigma)} \frac{\theta_j}{2} \geq (\sigma_i - 1)\frac{\theta_i}{2}$, which implies $u_i(\sigma_i, t_i; \theta_i) = -(\sigma_i - 1)\theta_i + \sum_{j\in P_i(\sigma)} \frac{\theta_j}{2} - (|N| - \sigma_i)\frac{\theta_i}{2} \geq -(|N| - 1)\frac{\theta_i}{2}$.

Next we show that φ^C satisfies *positive cost monotonicity*. Let $N \in \mathcal{N}$, θ, $\theta' \in \mathcal{Q}^N$, $(\sigma, t) \in \varphi^C(\theta)$, $(\sigma', t') \in \varphi^C(\theta')$, and $k \in N$ be such that for all $i \in N\backslash\{k\}$, $\theta_i = \theta'_i$ and $\theta_k < \theta'_k$. By *efficiency*, $\sigma_k \geq \sigma'_k$ and for all $i \in N\backslash\{k\}$, $\sigma_i > \sigma_k$ implies $\sigma'_i > \sigma'_k$. We partition $N\backslash\{k\}$ into N_1, N_2, N_3 defined by

$N_1 = \{i \in N | \sigma_k < \sigma_i\}$,
$N_2 = \{i \in N | \sigma_i < \sigma_k, \sigma'_k < \sigma'_i\}$, and
$N_3 = \{i \in N | \sigma'_i < \sigma'_k\}$.

We need to show that for all $i \in N\backslash\{k\}$, $u_i(\sigma_i, t_i; \theta_i) \leq u(\sigma'_i, t'_i; \theta'_i)$.

(i) If $i \in N_1$, then by *efficiency*, $\theta'_k \geq \theta_i$. Therefore, $u_i(\sigma'_i, t'_i; \theta_i) - u_i(\sigma_i, t_i; \theta_i) = \frac{\theta'_k}{2} - \frac{\theta_k}{2} > 0$, the desired conclusion.
(ii) If $i \in N_2$, then by *efficiency*, $\theta'_k \geq \theta_i$. Note that for all $j \in P_i(\sigma)\backslash P_i(\sigma')$ or $j \in P_i(\sigma')\backslash P_i(\sigma)$ such that $j \neq k$, $\theta_j = \theta_i$. Therefore,

$$\begin{aligned}
u_i(\sigma'_i, t'_i; \theta_i) - u_i(\sigma_i, t_i; \theta_i) &= -(|N|-1)\theta_i + \{\sum_{j \in P_i(\sigma')} \frac{\theta_j}{2} + (|N| - \sigma'_i)\frac{\theta_i}{2}\} \\
&\quad +(|N|-1)\theta_i - \{\sum_{j \in P_i(\sigma)} \frac{\theta_j}{2} + (|N| - \sigma_i)\frac{\theta_i}{2}\} \\
&= \frac{\theta'_k}{2} - \frac{\theta_i}{2} \\
&> 0,
\end{aligned}$$

the desired conclusion.
(iii) If $i \in N_3$, the desired conclusion comes from the fact that φ^C satisfies *independence of following costs*.

Now we show that φ^C satisfies *first-agent equal responsibility*. Let $N \in \mathcal{N}$, $\theta \in \mathcal{Q}^N$, and $(\sigma, t) \in \varphi^C(\theta)$. Suppose, without loss of generality, that $\sigma_1 = 1$. Then, $u_1(\sigma_1, t_1; \theta_1) = -(|N|-1)\frac{\theta_1}{2}$. Consider $\theta_{N\backslash\{1\}} \in \mathcal{Q}^{N\backslash\{1\}}$. By *efficiency*, for all $i \in N\backslash\{1\}$, there is $(\sigma', t') \in \varphi^C(\theta_{N\backslash\{1\}})$ such that $\sigma'_i = \sigma_i - 1$. Moreover, $t'_i - t_i = -\frac{\theta_1}{2}$, the desired conclusion.

Conversely, let φ be a rule satisfying the four axioms. Let $N \in \mathcal{N}$, $\theta \in \mathcal{Q}^N$, and $(\sigma, t) \in \varphi(\theta)$.

Step C-1: For $k \in N$ such that for all $i \in N$, $\theta_k \geq \theta_i$, $u_k(\sigma_k, t_k; \theta_k) = -\frac{|N|-1}{2}\theta_k$. By the *identical preferences lower bound*, $u_k(\sigma_k, t_k; \theta_k) \geq -\frac{|N|-1}{2}\theta_k$. Suppose, by way of contradiction, that the claim does not hold, that is, $u_k(\sigma_k, t_k; \theta_k) > -\frac{|N|-1}{2}\theta_k$. Let $\theta' \in \mathcal{Q}^N$ be such that for all $i \in N$, $\theta'_i = \theta_k$, and $(\sigma', t') \in \varphi(\theta')$. By repeated application of *positive cost monotonicity*, $u_k(\sigma'_k, t'_k; \theta_k) > -\frac{|N|-1}{2}\theta_k$. By the *identical preferences lower bound,* for all $i \in N\backslash\{k\}$, $u_i(\sigma'_i, t'_i; \theta'_i) \geq -\frac{|N|-1}{2}\theta_k$. Altogether,

$$\sum_{i \in N} u_i(\sigma'_i, t'_i; \theta'_i) > -\frac{|N|(|N|-1)}{2}\theta_k.$$

On the other hand, *feasibility* of a rule requires that

$$\sum_{i \in N} t_i = \sum_{i \in N} u_i(\sigma'_i, t'_i; \theta'_i) + \{1 + \cdots + (|N|-1)\}\theta_k \leq 0,$$

or equivalently,

$$\sum_{i \in N} u_i(\sigma'_i, t'_i; \theta'_i) \leq -\frac{|N|(|N|-1)}{2}\theta_k,$$

a contradiction.

Step C-2: If there are $k, \ell \in N$ such that for all $i \in N\backslash\{k\}$, $\theta_i \leq \theta_k$ and $\sigma_\ell = 1$, then $\theta_\ell = \theta_k$. From Step C-1, $u_k(\sigma_k, t_k; \theta_k) = -\frac{|N|-1}{2}\theta_k$. Suppose, by way of contradiction, that $\theta_\ell < \theta_k$. By *first-agent equal responsibility,* there is $(\sigma', t') \in \varphi(\theta_{N\backslash\{\ell\}})$ such that $u_k(\sigma'_k, t'_k; \theta_k) = -\frac{|N|-1}{2}\theta_k + \frac{t_\ell}{|N|-1} + \theta_k$. Note that for all $i \in N\backslash\{\ell\}$, $\theta_i \leq \theta_k$. From Step C-1, $u_k(\sigma'_k, t'_k; \theta_k) = -\frac{|N|-2}{2}\theta_k$, which implies $t_\ell = -\frac{|N|-1}{2}\theta_k$. Taken together, $u_\ell(\sigma_\ell, t_\ell; \theta_\ell) = -\frac{|N|-1}{2}\theta_k < -\frac{|N|-1}{2}\theta_\ell$, in violation of the *identical preferences lower bound.* Therefore, if $\sigma_\ell = 1$, then $\theta_\ell = \theta_k$ and $t_\ell = -\frac{|N|-1}{2}\theta_\ell$.

Step C-3: Now we show that all agents should end up with the utilities assigned by the maximal transfer rule, beginning with the agent in the first position in the queue. Let $\ell \in N$ be such that $\sigma_\ell = 1$. From Step C-2, $t_\ell = -\frac{|N|-1}{2}\theta_\ell$ and for all $i \in N\backslash\{k\}$, $\theta_i \leq \theta_\ell$. Let $L = N\backslash\{\ell\}$ and $k \in L$ be such that for all $i \in L\backslash\{k\}$, $\theta_i \leq \theta_k$. By *first-agent equal responsibility,* there is $(\sigma', t') \in \varphi(\theta_L)$ such that $\sigma'_k = \sigma_k - 1$. From Step C-1 applied to L and k, $u_k(\sigma'_k, t'_k; \theta_k) = -\frac{|L|-1}{2}\theta_k$. Moreover, by *first-agent equal responsibility,*

$$u_k(\sigma'_k, t'_k; \theta_k) = u_k(\sigma_k, t_k; \theta_k) - \frac{\theta_\ell}{2} + \theta_k,$$

so that

$$u_k(\sigma_k, t_k; \theta_k) = \frac{\theta_\ell}{2} - \frac{|N|-2}{2}\theta_k - \theta_k.$$

By repeating the argument, we obtain that for all $i \in N$,

$$u_i(\sigma_i, t_i; \theta_i) = \sum_{j \in P_i(\sigma)} \frac{\theta_j}{2} - \frac{|N| + \sigma_i - 2}{2}\theta_i,$$

which implies that $t_i = \sum_{j \in P_i(\sigma)} \frac{\theta_j}{2} - (|N| - \sigma_i)\frac{\theta_i}{2}$.

Step C-4: By *Pareto indifference,* we obtain the desired conclusion. □

Remark 4.1 The minimal transfer rule minimizes the sum of the absolute value of transfers among all rules satisfying *Pareto indifference,* the *identical preferences lower bound,* and *last-agent equal responsibility* (Maniquet 2003). On the other hand, the maximal transfer rule maximizes the sum of the absolute value of transfers among all rules satisfying *Pareto indifference,* the *identical preferences lower bound,* and *first-agent equal responsibility* (Chun 2006). We note that a small aggregate transfer in general is good news for the agents served earlier (the one receiving negative transfers), but bad news for the agents served later (the ones receiving positive transfers).

4.4 Balanced Consistency and Balanced Cost Reduction

In this section, we investigate how the minimal and the maximal transfer rules respond to changes in the set of agents. *Balanced consistency* requires that the effect on the payoff from the departure of one agent to another agent should be equal between any two agents. On the other hand, *balanced cost reduction* requires that if one agent leaves a problem, then the total payoffs of the remaining agents should be affected by the amount previously assigned to the leaving agent. The minimal transfer rule is the only rule satisfying *efficiency* and *Pareto indifference* together with either one of the two axioms, *balanced consistency* and *balanced cost reduction.* On the other hand, the maximal transfer rule satisfies an alternative formulation of *balanced consistency under constant completion time*: upon the departure of an agent, all of her predecessors are assumed to move back by one position to keep the completion time constant. Under this alternative formulation, the maximal transfer rule becomes the only rule satisfying *efficiency, Pareto indifference,* and *balanced consistency under constant completion time.*

We note that the minimal and the maximal transfer rules assign a unique allocation if and only if all agents have different waiting costs. However, even when some agents have the same waiting cost, agents' utilities do not depend on the choice of efficient queues if the compensation is determined according to the minimal or the maximal transfer rule. Thus, both rules are *essentially single-valued,* in the sense that for a given problem, each agent's utility is the same at all allocations that the rule chooses. As a consequence, any efficient queue can be chosen to calculate the utilities assigned by the two rules. In the formulation of our next two axioms, *balanced consistency* and *balanced cost reduction,* we assume that a rule is *essentially single-valued.*

If an agent leaves a queueing problem, then it will affect the payoffs of other remaining agents. *Balanced consistency* requires that the effect of agent i leaving a queueing problem on the payoff of another agent $j \neq i$ should be the same as the effect of agent j leaving a queueing problem on the payoff of agent i. It is similar to "preservation of differences" of solutions for games (Hart and Mas-Colell 1989).[3] To stress the fact that the axiom concerns situations in which an agent leaves a queueing problem similar as players leave a game in consistency properties, this property is referred to as *balanced consistency*.[4]

Balanced consistency: For all $N \in \mathcal{N}$, all $\theta \in \mathcal{Q}^N$, all $i, j \in N$, all $(\sigma, t) \in \varphi(\theta)$, all $(\sigma^{-i}, t^{-i}) \in \varphi(\theta_{N\setminus\{i\}})$, and all $(\sigma^{-j}, t^{-j}) \in \varphi(\theta_{N\setminus\{j\}})$,

$$u_i(\sigma, t) - u_i(\sigma^{-j}, t^{-j}) = u_j(\sigma, t) - u_j(\sigma^{-i}, t^{-i}).$$

Now we investigate the implications of *balanced consistency* in the context of queueing problems. First, we show that the minimal transfer rule satisfies the property.

Lemma 4.1 *The minimal transfer rule satisfies balanced consistency.*

Proof Let $N \in \mathcal{N}$ and $\theta \in \mathcal{Q}^N$. From the *essential single-valuedness* of φ^M, we may choose any $(\sigma, t) \in \varphi^M(\theta)$. Let $i, j \in N$ be such that $j \in P_i(\sigma)$ (and thus $i \in F_j(\sigma)$). To simplify the notation, we do not attach the superscript M to σ and t. Then, for all $(\sigma^{-i}, t^{-i}) \in \varphi^M(\theta_{N\setminus\{i\}})$ and all $(\sigma^{-j}, t^{-j}) \in \varphi^M(\theta_{N\setminus\{j\}})$,

$$\begin{aligned} & u_i(\sigma, t) - u_i(\sigma^{-j}, t^{-j}) \\ & = -(\sigma_i - 1)\frac{\theta_i}{2} - \sum_{k \in F_i(\sigma)} \frac{\theta_k}{2} - \left(-(\sigma_i - 2)\frac{\theta_i}{2} - \sum_{k \in F_i(\sigma)} \frac{\theta_k}{2} \right) \\ & = -\frac{\theta_i}{2}, \end{aligned} \tag{4.1}$$

and

$$\begin{aligned} & u_j(\sigma, t) - u_j(\sigma^{-i}, t^{-i}) \\ & = -(\sigma_j - 1)\frac{\theta_j}{2} - \sum_{k \in F_j(\sigma)} \frac{\theta_k}{2} - \left(-(\sigma_j - 1)\frac{\theta_j}{2} - \sum_{k \in F_j(\sigma)\setminus\{i\}} \frac{\theta_k}{2} \right) \\ & = -\frac{\theta_i}{2}. \end{aligned} \tag{4.2}$$

[3]This property states that the effect of player i leaving the game on the payoff of player $j \neq i$ is equal to the effect of player j leaving the game on the payoff of player i.

[4]Note that in the *balanced contributions property* introduced by Myerson (1980) for cooperative games with a restricted set of feasible coalitions, the player set is fixed.

Altogether, we conclude that the minimal transfer rule satisfies *balanced consistency.* □

We ask whether there is any other rule satisfying *efficiency* and *Pareto indifference* together with *balanced consistency.* As it turns out, the minimal transfer rule is the only one satisfying the three axioms together. Before proceeding to our first main result, we state the following lemma.

Lemma 4.2 *If a rule φ is efficient, then for all $N \in \mathcal{N}$, all $\theta \in \mathcal{Q}^N$, all $(\sigma, t) \in \varphi(\theta)$, all $i \in N$, and all $(\sigma^{-i}, t^{-i}) \in \varphi(\theta_{N\setminus\{i\}})$,*

$$\sum_{j\in N} u_j(\sigma, t) - \sum_{j\in N\setminus\{i\}} u_j(\sigma^{-i}, t^{-i}) = -(\sigma_i - 1)\theta_i - \sum_{k\in F_i(\sigma)} \theta_k. \tag{4.3}$$

Proof For all $(\sigma, t) \in \varphi(\theta)$, all $i \in N$, and all $(\sigma^{-i}, t^{-i}) \in \varphi(\theta_{N\setminus\{i\}})$, *efficiency* implies that

$$\sum_{j\in N} u_j(\sigma, t) = -\sum_{j\in N}(\sigma_j - 1)\theta_j$$

and

$$\sum_{j\in N\setminus\{i\}} u_j(\sigma^{-i}, t^{-i}) = -\sum_{j\in P_i(\sigma)} (\sigma_j - 1)\theta_j - \sum_{j\in F_i(\sigma)} \sum_{k\in P_j(\sigma)\setminus\{i\}} \theta_j.$$

Since, by efficiency, all agents are served in the nonincreasing order with respect to their waiting costs, subtracting these two equations from each other, we obtain Eq. (4.3). □

Now we are ready to show our characterization result based on *balanced consistency*. We note that if $\sigma_i \in \{1, \ldots, |N|\}$ is determined and $u_i(\sigma, t)$ is known, then also t_i is determined.

Theorem 4.3 (van den Brink and Chun 2012) *The minimal transfer rule is the only rule satisfying efficiency, Pareto indifference, and balanced consistency.*

Proof It is well-known that the minimal transfer rule satisfies *efficiency* and *Pareto indifference*, and from Lemma 4.1, it satisfies *balanced consistency.* Conversely, let φ be a rule satisfying the three axioms. Let $N \in \mathcal{N}$ and $\theta \in \mathcal{Q}^N$ be given. If $|N| = 1$, then *efficiency* implies that $\sigma_i = 1$ and $t_i = 0$ for $i \in N$.

Let N be such that $|N| = 2$. Without loss of generality, we may assume that $N = \{i, j\}$ and that $\theta_i \geq \theta_j$. Let $(\sigma, t) \in \varphi(\theta)$, $(\sigma^{-i}, t^{-i}) \in \varphi(\theta_{N\setminus\{i\}})$, and $(\sigma^{-j}, t^{-j}) \in \varphi(\theta_{N\setminus\{j\}})$. By *balanced consistency,* $u_i(\sigma, t) - u_i(\sigma^{-j}, t^{-j}) = u_j(\sigma, t) - u_j(\sigma^{-i}, t^{-i})$. Since $u_i(\sigma^{-j}, t^{-j}) = u_j(\sigma^{-i}, t^{-i}) = 0$, we have $u_i(\sigma, t) = u_j(\sigma, t)$. By *efficiency,* $u_i(\sigma, t) + u_j(\sigma, t) = -\theta_j$. Altogether, we obtain $u_i(\sigma, t) = u_j(\sigma, t) = -\frac{\theta_j}{2}$. By *efficiency* and *Pareto indifference,* we may assume that $\sigma_i = 1$ and $\sigma_j = 2$. Then, $t_i = -\frac{\theta_j}{2} = -t_j$, as desired.

We will establish the claim for an arbitrary number of agents by an induction argument. Let $N = \{1, 2, \ldots, n\}$, $\theta \in \mathscr{Q}^N$ and $(\sigma, t) \in \varphi(\theta)$. By *efficiency* and *Pareto indifference,* we may assume without loss of generality that $\theta_1 \geq \theta_2 \geq \cdots \geq \theta_n$, and that $\sigma_i = i$ for all $i \in N$. As induction hypothesis, suppose that for all $N' \in \mathscr{N}$ such that $|N'| \leq |N| - 1$ and all $\theta' \in \mathscr{Q}^{N'}$, $\varphi(\theta') = \varphi^M(\theta')$. For all $i \in N$, let $(\sigma^{-i}, t^{-i}) \in \varphi(\theta_{N\setminus\{i\}})$. By *balanced consistency,* for all $i, j \in N$, $u_i(\sigma, t) - u_i(\sigma^{-j}, t^{-j}) = u_j(\sigma, t) - u_j(\sigma^{-i}, t^{-i})$. Now fix i, change $j \neq i$ from 1 to n, and add up the $(n-1)$ equations obtained in this way. We have:

$$(n-1)u_i(\sigma, t) - \sum_{j\in N\setminus\{i\}} u_i(\sigma^{-j}, t^{-j}) = \sum_{j\in N\setminus\{i\}} (u_j(\sigma, t) - u_j(\sigma^{-i}, t^{-i})).$$

Adding $u_i(\sigma, t) + \sum_{j\in N\setminus\{i\}} u_i(\sigma^{-j}, t^{-j})$ to both sides gives

$$n \cdot u_i(\sigma, t) = \sum_{j\in N} u_j(\sigma, t) - \sum_{j\in N\setminus\{i\}} u_j(\sigma^{-i}, t^{-i}) + \sum_{j\in N\setminus\{i\}} u_i(\sigma^{-j}, t^{-j}). \tag{4.4}$$

From the induction hypothesis, it follows that

$$\begin{aligned}
\sum_{j\in N\setminus\{i\}} u_i(\sigma^{-j}, t^{-j}) &= -\sum_{j\in P_i(\sigma)} \left(\sum_{k\in P_i(\sigma)\setminus\{j\}} \frac{\theta_i}{2} - \sum_{k\in F_i(\sigma)} \frac{\theta_k}{2} \right) \\
&\quad + \sum_{j\in F_i(\sigma)} \left(\sum_{k\in P_i(\sigma)} \frac{\theta_i}{2} - \sum_{k\in F_i(\sigma)\setminus\{j\}} \frac{\theta_k}{2} \right) \\
&= -(i-2)(\sigma_i - 1)\frac{\theta_i}{2} - (i-1) \sum_{k\in F_i(\sigma)} \frac{\theta_k}{2} \\
&\quad -(n-i)(\sigma_i - 1)\frac{\theta_i}{2} - (n-i-1) \sum_{k\in F_i(\sigma)} \frac{\theta_k}{2} \\
&= -(n-2)(\sigma_i - 1)\frac{\theta_i}{2} - (n-2) \sum_{k\in F_i(\sigma)} \frac{\theta_k}{2}.
\end{aligned} \tag{4.5}$$

Using Lemma 4.2, substituting Eqs. (4.3) and (4.5) into Eq. (4.4) yields

$$n \cdot u_i(\sigma, t) = -2(\sigma_i - 1)\frac{\theta_i}{2} - 2 \sum_{k\in F_i(\sigma)} \frac{\theta_k}{2}$$

$$-(n-2)(\sigma_i - 1)\frac{\theta_i}{2} - (n-2) \sum_{k \in F_i(\sigma)} \frac{\theta_k}{2}$$

$$= -n(\sigma_i - 1)\frac{\theta_i}{2} - n \sum_{k \in F_i(\sigma)} \frac{\theta_k}{2},$$

or equivalently,

$$u_i(\sigma, t) = -(\sigma_i - 1)\frac{\theta_i}{2} - \sum_{k \in F_i(\sigma)} \frac{\theta_k}{2} = u_i(\sigma^M, t^M). \tag{4.6}$$

By *efficiency*, $\sigma \in Eff(\theta)$, and thus Eq. (4.6) fixes the transfers

$$t_i = (\sigma_i - 1)\frac{\theta_i}{2} - \sum_{j \in F_i(\sigma)} \frac{\theta_j}{2} = t_i^M,$$

as desired. □

Remark 4.2 Although "preservation of differences" is typical for the Shapley value, it is not obvious for queueing problems that *balanced consistency* characterizes the minimal transfer rule since the maximal transfer rule is also obtained as the Shapley value of a pessimistic queueing game (for details, see Sect. 3.4).

Remark 4.3 Upon the departure of an agent, if we assume that all of her predecessors are moving back by one position to keep the same completion time, an alternative *balanced consistency under constant completion time* property can be formulated. The maximal transfer rule is the only rule satisfying *efficiency, Pareto indifference,* and *balanced consistency under constant completion time.*

Now suppose that an agent leaves a queueing problem. Since the agent is not in the queue anymore, the total waiting cost of all the remaining agents will be decreased. In other words, the presence of an agent generates a negative externality to any other agent. *Balanced cost reduction* requires that the total (overall remaining agents) decrease in this negative externality as a result of the departure of an agent be equal to the negative of the payoff of the departing agent when she is still present.

Balanced cost reduction: For all $N \in \mathcal{N}$, all $\theta \in \mathcal{Q}^N$, all $j \in N$, all $(\sigma, t) \in \varphi(\theta)$, and all $(\sigma^{-j}, t^{-j}) \in \varphi(\theta_{N\setminus\{j\}})$,

$$\sum_{i \in N\setminus\{j\}} \left(u_i(\sigma, t) - u_i(\sigma^{-j}, t^{-j})\right) = u_j(\sigma, t).$$

We explore the implications of *balanced cost reduction* in the context of queueing problems. First, we show that the minimal transfer rule satisfies the property.

Lemma 4.3 *The minimal transfer rule satisfies balanced cost reduction.*

Proof Let $N \in \mathcal{N}$, $\theta \in \mathcal{Q}^N$ and $j \in N$. From the *essential single-valuedness* of φ^M, we may choose any $(\sigma, t) \in \varphi^M(\theta)$ and any $(\sigma^{-j}, t^{-j}) \in \varphi^M(\theta_{N\setminus\{j\}})$. To simplify the notation, we do not attach the superscript M to σ and t. From Eqs. (4.1) and (4.2) in Lemma 4.1, we can derive that

$$\begin{aligned}\sum_{i \in N\setminus\{j\}} (u_i(\sigma, t) - u_i(\sigma^{-j}, t^{-j})) &= -(\sigma_j - 1)\frac{\theta_j}{2} - \sum_{i \in F_j(\sigma)} \frac{\theta_i}{2} \\ &= u_j(\sigma, t),\end{aligned}$$

showing that the minimal transfer rule satisfies *balanced cost reduction.* □

Next we present our second characterization.

Theorem 4.4 (van den Brink and Chun 2012) *The minimal transfer rule is the only rule satisfying efficiency, Pareto indifference, and balanced cost reduction.*

Proof It is well-known that the minimal transfer rule satisfies *efficiency* and *Pareto indifference*, and from Lemma 4.3, it satisfies *balanced cost reduction.*

Conversely, let φ be a rule satisfying the three axioms. Let $N \in \mathcal{N}$ and $\theta \in \mathcal{Q}^N$ be given. If $|N| = 1$, then *efficiency* implies that $\sigma_i = 1$ and $t_i = 0$ for $i \in N$. Now let N be such that $|N| \geq 2$, $j \in N$ be a leaving agent, and $(\sigma^{-j}, t^{-j}) \in \varphi(\theta_{N\setminus\{j\}})$. By *balanced cost reduction,*

$$\sum_{i \in N\setminus\{j\}} \left(u_i(\sigma, t) - u_i(\sigma^{-j}, t^{-j})\right) = u_j(\sigma, t).$$

Adding $u_j(\sigma, t)$ to both sides gives

$$\sum_{i \in N} u_i(\sigma, t) - \sum_{i \in N\setminus\{j\}} u_i(\sigma^{-j}, t^{-j}) = 2u_j(\sigma, t). \tag{4.7}$$

Using Lemma 4.2 and substituting Eq. (4.3) into Eq. (4.7) yield

$$2u_j(\sigma, t) = -(\sigma_j - 1)\theta_j - \sum_{i \in F_j(\sigma)} \theta_i,$$

which implies that

$$u_j(\sigma, t) = -(\sigma_j - 1)\frac{\theta_j}{2} - \sum_{i \in F_j(\sigma)} \frac{\theta_i}{2} = u_j(\sigma^M, t^M). \tag{4.8}$$

By *efficiency* and *Pareto indifference,* we choose all efficient queues in $Eff(\theta)$, and for any $\sigma \in Eff(\theta)$, Eq. (4.8) fixes the transfers

$$t_j = (\sigma_j - 1)\frac{\theta_j}{2} - \sum_{i \in F_j(\sigma)} \frac{\theta_i}{2} = t_j^M,$$

the desired expression.[5] □

Remark 4.4 We note that our two characterizations of the minimal transfer rule on the basis of *balanced consistency* or *balanced cost reduction* in addition to *efficiency* and *Pareto indifference* carry over to either sequencing problems or scheduling problems. In fact, van den Brink and Chun (2012) present their characterizations in the context of sequencing problems.

4.5 Concluding Remarks

In this chapter, we present characterizations of the minimal and the maximal transfer rules on the basis of various axioms. Another axiom widely discussed in the literature specifying how a rule should respond to changes in the population is *population solidarity* (Chun 1986; Thomson 1983a,b; and others[6]): it requires that upon the departure of an agent, all the remaining agents should be affected in the same direction, all gain or all lose. The minimal transfer rule satisfies *population solidarity,* but the maximal transfer rule does not satisfy it (Chun 2006). On the other hand, as in Remark 4.3, upon the departure of an agent, if we assume that all of her predecessors are moving back by one position to keep the completion time constant, then both the minimal and the maximal transfer rules satisfy the alternative *population solidarity under constant completion time* property. It remains an open question whether the minimal or the maximal transfer rules can be characterized on the basis of *population solidarity.*

Another question for future research is to investigate axioms concerning changes in the parameters of the queueing problem without changing the set of agents, such as the before mentioned *balanced contributions* property (Myerson 1980) or *fairness* (Myerson 1977; van den Brink 2001).

[5]Note the difference with the end of the proof of Theorem 4.3. In the proof of Theorem 4.4, we need to make sure to choose all efficient queues, and for any efficient queue, we have the desired formula. In the induction proof of Theorem 4.3, one efficient queue is chosen already from the beginning.

[6]See Thomson (1995) for a survey.

References

Chun, Y. (1986). The solidarity axiom for quasi-linear social choice problems. *Social Choice and Welfare, 3*, 297–310.

Chun, Y. (2006). A pessimistic approach to the queueing problem. *Mathematical Social Sciences, 51*, 171–181.

Chun, Y., & Thomson, W. (1988). Monotonicity properties of bargaining solutions when applied to economies. *Mathematical Social Sciences, 15*, 11–27.

Hart, S., & Mas-Colell, A. (1989). Potential, value and consistency. *Econometrica, 57*, 589–614.

Maniquet, F. (2003). A characterization of the Shapley value in queueing problems. *Journal of Economic Theory, 109*, 90–103.

Moulin, H. (1987). The pure compensation problem: Egalitarianism versus laissez-fairism. *Quarterly Journal of Economics, 102*, 769–783.

Moulin, H., & Thomson, W. (1988). Can everyone benefit from growth? Two difficulties. *Journal of Mathematical Economics, 17*, 339–345.

Myerson, R. B. (1977). Graphs and cooperation in games. *Mathematics of Operations Research, 2*, 225–229.

Myerson, R. B. (1980). Conference structures and fair allocation rules. *International Journal of Game Theory, 9*, 169–182.

Roemer, J. (1986). Equality of resources implies equality of welfare. *Quarterly Journal of Economics, 101*, 751–784.

Thomson, W. (1983a). The fair division of a fixed supply among a growing population. *Mathematics of Operations Research, 8*, 319–326.

Thomson, W. (1983b). Problems of fair division and the Egalitarian solution. *Journal of Economic Theory, 31*, 211–226.

Thomson, W. (1995). Population monotonic allocation rules. In W. A. Barnett, H. Moulin, M. Salles & N. J. Schofield (Eds.), *Social choice, welfare, and ethics* (pp. 79–124). Cambridge: Cambridge University Press.

Thomson, W. (1999). Welfare-domination under preference-replacement: A survey and open questions. *Social Choice and Welfare, 16*, 373–394.

van den Brink, R. (2001). An axiomatization of the Shapley value using a fairness property. *International Journal of Game Theory, 30*, 309–319.

van den Brink, R., & Chun, Y. (2012). Balanced consistency and balanced cost reduction for sequencing problems. *Social Choice and Welfare, 38*, 519–529.

Chapter 5
No-Envy

5.1 Introduction

No-envy, introduced by Foley (1967), requires that no agent should end up with a higher utility by consuming what any other agent consumes, and its implications have been studied for a wide class of problems. In this chapter, we investigate the implications of *no-envy* in the context of queueing problems.

First, it is not difficult to show that *no-envy* implies *queue-efficiency*. Then, we identify an easy way of checking whether a rule satisfies *no-envy*. It can be described in a simple way: choose any efficient queue, and then check the difference of transfers between any two neighboring agents. If the difference is not greater than the higher waiting cost of the two agents and is not smaller than the lower waiting cost of the two agents, then it passes the *no-envy* test. Of course, it is an immediate consequence of *no-envy* that an agent served earlier should receive a smaller transfer than an agent served later. The existence of such a rule can easily be established.

We also investigate whether there is a rule satisfying *efficiency* and *no-envy* together with either one of two cost monotonicity axioms, *negative cost monotonicity* and *positive cost monotonicity* (see Sect. 4.3 for more explanation on these axioms). We show that if the society consists of more than two agents, then there is no rule satisfying *efficiency, no-envy,* and either *negative cost monotonicity* or *positive cost monotonicity.*[1]

Faced with the impossibility results, we propose modifications of *no-envy*. To apply *no-envy,* each agent is supposed to reevaluate what any other agents consume. In the queueing problem, an allocation consists of agents' positions in the queue and their transfers, and a rule determines the transfers by agents' positions and

[1]As we show later, if the society consists of only two agents, then the minimal transfer rule satisfies *efficiency, no-envy,* and *negative cost monotonicity,* and the maximal transfer rule satisfies *efficiency, no-envy,* and *positive cost monotonicity.* Moreover, the rules can be characterized by these axioms if *Pareto indifference* is additionally imposed. See Remark 5.2 for details.

© Springer International Publishing Switzerland 2016

Y. Chun, *Fair Queueing*, Studies in Choice and Welfare,
DOI 10.1007/978-3-319-33771-5_5

waiting costs. If two agents interchange their positions in the queue, then their transfers would not be the same as before because their waiting costs are different. *Adjusted no-envy* requires that an agent should not envy the other agents after making the adjustment in transfers. Our second modification of *no-envy* requires that an agent should not envy the other agents at least in one direction. More specifically, *backward no-envy* requires that an agent should not envy the agents with lower waiting costs, whereas *forward no-envy* requires that an agent should not envy the agents with higher waiting costs. For each of these modifications, we show that the impossibility results do not hold any more. In fact, the minimal transfer rule satisfies *efficiency, negative cost monotonicity, adjusted no-envy,* and *backward no-envy,* and the maximal transfer rule satisfies *efficiency, positive cost monotonicity, adjusted no-envy,* and *forward no-envy.*

Other fairness requirements widely discussed in the literature are the *identical preferences lower bound* which requires that each agent should be at least as well off as she would be, under *efficiency* and *equal treatment of equals,* if all other agents had the same preferences, and *egalitarian equivalence* which requires that there should be a reference bundle such that each agent enjoys the same welfare between her bundle and that reference bundle. We investigate whether the three requirements are compatible in the current context. First, it is easy to show that *efficiency* and *no-envy* together imply the *identical preferences lower bound.* Also, we can show an existence of a rule satisfying *efficiency, egalitarian equivalence,* and the *identical preferences lower bound.* However, if we have more than three agents, then there is no rule satisfying *no-envy* and *egalitarian equivalence* together.

The chapter is organized as follows.[2] Section 5.2 explores the implications of *no-envy,* and Sect. 5.3 shows that the incompatibility of *efficiency, no-envy,* and either one of two *cost monotonicity* axioms. Section 5.4 introduces two modifications of *no-envy, adjusted no-envy* and *backward/forward no-envy,* and investigates their implications. Section 5.5 discusses whether three fairness requirements, *no-envy,* the *identical preferences lower bound,* and *egalitarian equivalence,* are compatible in this context. Concluding remarks follow in Sect. 5.6.

5.2 Efficiency and No-Envy

No-envy requires that no agent should end up with a higher utility by consuming what any other agent consumes. It is a standard requirement in the studies of fairness for a wide class of problems (Thomson 2005; Thomson and Varian 1985). Given $N \in \mathcal{N}$ and $\theta \in \mathcal{Q}^N$, an allocation $(\sigma, t) \in Z(\theta)$ satisfies *no-envy* if for all $i, j \in N$, $u_i(\sigma_i, t_i; \theta_i) \geq u_i(\sigma_j, t_j; \theta_i)$. Let $F(\theta)$ be the set of all no-envy allocations for $\theta \in \mathcal{Q}^N$.

No-envy: For all $N \in \mathcal{N}$, all $\theta \in \mathcal{Q}^N$, and all $(\sigma, t) \in \varphi(\theta)$, $(\sigma, t) \in F(\theta)$.

[2]This chapter is based mainly on Chun (2006). Propositions 5.1 and 5.4 are based on Chun et al. (2014).

It is not difficult to show that *no-envy* implies *queue-efficiency,* which requires that agents should be served in the non-increasing order with respect to their unit waiting costs.

Proposition 5.1 (Chun et al. 2014) *No-envy implies queue-efficiency.*

Proof Let φ be a rule satisfying *no-envy*. Let $N \in \mathcal{N}$, $\theta \in \mathcal{Q}^N$, $i, j \in N$, and $(\sigma, t) \in \varphi(\theta)$. For agent i not to envy j at (σ, t), we have

$$-(\sigma_i - 1)\theta_i + t_i \geq -(\sigma_j - 1)\theta_i + t_j \quad \text{or} \quad t_i - t_j \geq (\sigma_i - \sigma_j)\theta_i.$$

On the other hand, for agent j not to envy agent i, we have

$$-(\sigma_j - 1)\theta_j + t_j \geq -(\sigma_i - 1)\theta_j + t_i \quad \text{or} \quad t_j - t_i \geq (\sigma_j - \sigma_i)\theta_j.$$

Combining these two inequalities together, we obtain

$$(\sigma_i - \sigma_j)\theta_i \leq t_i - t_j \leq (\sigma_i - \sigma_j)\theta_j. \tag{5.1}$$

The two conditions on the transfers are compatible only when $(\sigma_i - \sigma_j)(\theta_i - \theta_j) \leq 0$. Hence, if $\theta_i > \theta_j$, then $\sigma_i < \sigma_j$,[3] which is the condition for *queue-efficiency*. □

Now we present a simple way of checking whether a rule satisfies *no-envy.*

Theorem 5.1 *A rule φ satisfies no-envy if and only if for all $N \in \mathcal{N}$ such that $N = \{1, \ldots, n\}$, all $\theta \in \mathcal{Q}^N$ and all $(\sigma, t) \in \varphi(\theta)$, $\sigma \in Eff(\theta)$ and for all $\sigma_i = 1, \ldots, n-1$, $\theta_{\sigma_i} \geq t_{\sigma_i+1} - t_{\sigma_i} \geq \theta_{\sigma_i+1}$.*

Proof Let φ be a rule satisfying *no-envy*. Let $N = \{1, \ldots, n\}$, $\theta \in \mathcal{Q}^N$, and $(\sigma, t) \in \varphi(\theta)$. By Proposition 5.1, since *no-envy* implies *queue-efficiency*, $\sigma \in Eff(\theta)$. To simplify notation, we assume that $\theta_1 \geq \theta_2 \geq \cdots \geq \theta_n$, and for all $i \in N$, $\sigma_i = i$. Let $i, j \in N$. We may also assume, without loss of generality, that $j = i + k$ for $k \in \mathbf{N}$.

First, for i not to envy j, $u_i(\sigma_i, t_i; \theta_i) \geq u_i(\sigma_j, t_j; \theta_i)$, which is equivalent to $t_i - (i-1)\theta_i \geq t_j - (j-1)\theta_i$ or $k\theta_i \geq t_j - t_i$. In particular, if $j = i + 1$, this inequality becomes $\theta_i \geq t_{i+1} - t_i$, the desired expression. Applying this inequality recursively, for all $\ell = i, \ldots, i+k-1$, we have $\theta_\ell \geq t_{\ell+1} - t_\ell$. Summing these inequalities, $\sum_{\ell=i}^{i+k-1} \theta_\ell \geq t_{i+k} - t_i$. Since *queue-efficiency* (implied by *no-envy*) implies that $k\theta_i \geq \sum_{\ell=i}^{i+k-1} \theta_\ell$, we have $k\theta_i \geq t_j - t_i$. Therefore, it is sufficient to check the inequality between neighboring agents.

Similarly, for j not to envy i, $t_j - t_i \geq k\theta_{i+1}$. In particular, if $j = i + 1$, this inequality becomes $t_{i+1} - t_i \geq \theta_{i+1}$. Once again, by the same reasoning, we obtain the desired conclusion. □

Example 5.1 (The Minimal and the Maximal Transfer Rules Do Not Satisfy No-Envy) Let $N \in \mathcal{N}$ be such that $N = \{1, 2, 3\}$, $\theta \in \mathcal{Q}^N$, and $\theta = (6, 4, 2)$. Then,

[3] Since this is a queue on a single machine, $\sigma_i \neq \sigma_j$.

$\varphi^M(\theta) = (\sigma^M, t^M)$ is obtained by setting $\sigma^M = (1, 2, 3)$ and $t^M = (-3, 1, 2)$. Note that agent 3 envies agent 2 since $u(\sigma_2^M, t_2^M; \theta_3) = -1 > u(\sigma_3^M, t_3^M; \theta_3) = -2$. On the other hand, $\varphi^C(\theta) = (\sigma^C, t^C)$ is obtained by setting $\sigma^C = (1, 2, 3)$ and $t^C = (-6, 1, 5)$. Note that agent 1 envies agent 2 since $u(\sigma_2^C, t_2^C; \theta_1) = -5 > u(\sigma_1^C, t_1^C; \theta_1) = -6$.

Example 5.2 (Rules Satisfying Efficiency and No-Envy) Let $N \in \mathcal{N}$ be such that $N = \{1 \ldots, n\}$ and $\theta \in \mathcal{Q}^N$. To simplify our notation, we suppose that $\theta_1 \geq \cdots \geq \theta_n$ and $\sigma_i = i$. First, let $t_1 = \alpha_1$. We will determine α_1 after considering the budget constraint. Now, for $i = 2, \ldots, n$, we choose $\alpha_i \in [\theta_i, \theta_{i-1}]$ and $t_i = \sum_{j=1}^{i} \alpha_j$. Finally, we choose α_1 such that $\sum_{i \in N} t_j = 0$. An alternative rule can be given starting from n. Once again, let $t_n = -\beta_n$. We will determine β_n after considering the budget constraint. Now, for $i = n-1, \ldots, 1$, we choose $\beta_i \in [\theta_{i+1}, \theta_i]$ and $t_i = -\sum_{j=i}^{n} \beta_j$. Once again, we choose β_n such that $\sum_{i \in N} t_j = 0$. Clearly, these processes lead to rules satisfying *efficiency* and *no-envy*. Moreover, it is interesting to note that the symmetrically balanced VCG rule satisfies *efficiency* and *no-envy* in the current context (Chun 2005).

Remark 5.1 As shown in Svensson (1983), in economies with indivisible goods, *no-envy* implies *object-efficiency*.[4] Also, *no-envy* is equivalent to *group no-envy*,[5] and the set of *envy-free* allocations coincides with the set of *equal income Walrasian allocations*.[6] Similar observations can be made for queueing problems.[7]

5.3 No-Envy and Cost Monotonicity

We investigate whether there is a rule satisfying *efficiency* and *no-envy* together with either one of two cost monotonicity axioms (Sect. 4.3): *negative cost monotonicity* requires that an increase in an agent's waiting cost should cause all other agents to weakly lose, whereas *positive cost monotonicity* requires that an increase in an agent's waiting cost should cause all other agents to weakly gain. The answer is no.

Theorem 5.2 *Let $|N| \geq 3$. Then, there is no rule satisfying efficiency, no-envy, and either negative or positive cost monotonicity.*

---

[4]*Object-efficiency* requires that there is no feasible allocation which makes every agent better off and at least one agent strictly better off.

[5]Given two groups of the same size, suppose that a group redistributes among its members what is available to the other group. If a rule selects an allocation which is impossible to make every agent in the group better off, with at least one agent strictly better off, even after considering the possibility of redistribution, then the rule satisfies *group no-envy*.

[6]Allocations that can be supported as Walrasian equilibrium with an equal implicit income.

[7]For this, a position in a queue is considered as an indivisible good.

Proof

(i) Let φ be a rule satisfying *efficiency, no-envy,* and *negative cost monotonicity.* Let $N \in \mathcal{N}, \theta \in \mathcal{Q}^N$, and $(\sigma, t) \in \varphi(\theta)$ be such that $N = \{1, \dots, n\}$ with $n \geq 3$ and $\theta_1 > \theta_2 > \cdots > \theta_n$. By *queue-efficiency,* for all $i \in N, \sigma(i) = i$. Moreover, from Theorem 5.1, for all $i = 1, \dots, n-1, \theta_i \geq t_{i+1} - t_i \geq \theta_{i+1}$.

Case 1-1: There exists $i \in N \backslash \{n\}$ such that $\theta_i \geq t_{i+1} - \theta_{i+1}$. Let $\alpha \in]0, 1]$ be such that $t_{i+1} - t_i = \alpha\theta_i + (1-\alpha)(1-\alpha)\theta_{i+1} = \theta_{i+1} + \alpha(\theta_i - \theta_{i+1})$. Let $\varepsilon > 0$ be such that $\varepsilon < \alpha(\theta_i - \theta_{i+1})$ and $\theta'_i = \theta_{i+1} + \varepsilon$. Let θ' be the waiting cost vector obtained from θ by replacing θ_i with θ'_i and $(\sigma', t') \in \varphi(\theta')$.

By *negative cost monotonicity,* all agents except i weakly gain. Since the decrease of θ_i to θ'_i does not change the efficient queue, that is, for all $i \in N$, $\sigma'(i) = i$, this is possible only if for all $j \neq i, t'_j \geq t_j$. By *budget balance,* $t'_i \leq t_i$. Altogether, $t'_{i+1} - t'_i \geq t_{i+1} - t_i = \theta_{i+1} + \alpha(\theta_i - \theta_{i+1}) > \theta_{i+1} + \varepsilon = \theta'_i > \theta_{i+1}$, which contradicts the conclusion of Theorem 5.1.

Case 1-2: For all $i = 1, \dots, n-1, t_{i+1} - t_i = \theta_{i+1}$. Let $i \in N \backslash \{1, n\}$ and θ'_i be such that $\theta_i > \theta'_i > \theta_{i+1}$. Let θ' be the waiting cost vector obtained from θ by replacing θ_i with θ'_i and $(\sigma', t') \in \varphi(\theta')$. By *negative cost monotonicity,* all agents except i weakly gain. Since the decrease of θ_i to θ'_i does not change the efficient queue, this is possible only if for all $j \neq i, t'_j \geq t_j$. By *budget balance,* $t'_i \leq t_i$. Altogether, $t'_{i+1} - t'_i \geq t_{i+1} - t_i = \theta_{i+1}$. If $t'_{i+1} - t'_i > \theta_{i+1} = \theta'_{i+1}$, then we go back to Case 1-1 and obtain the desired conclusion. If $t'_{i+1} - t'_i = \theta_{i+1}$, by *budget balance,* we deduce that for all $j \in N, t'_j = t_j$. In particular, $t'_i - t'_{i-1} = \theta_i$. Since $\theta_{i-1} = \theta'_{i-1} > t'_i - t'_{i-1} > \theta'_i$, we go back to Case 1-1 and obtain the desired conclusion.

(ii) Let φ be a rule satisfying *efficiency, no-envy,* and *positive cost monotonicity.* Let $N \in \mathcal{N}, \theta \in \mathcal{Q}^N$, and $(\sigma, t) \in \varphi(\theta)$ be such that $N = \{1, \dots, n\}$ with $n \geq 3$ and $\theta_1 > \theta_2 > \cdots > \theta_n$. By *queue-efficiency,* for all $i \in N, \sigma(i) = i$. Moreover, from Theorem 5.1, for all $i = 1, \dots, n-1, \theta_i \geq t_{i+1} - t_i \geq \theta_{i+1}$.

Case 2-1: There exists $i \in N \backslash \{n\}$ such that $\theta_i > t_{i+1} - t_i \geq \theta_{i+1}$. Let $\alpha \in [0, 1[$ be such that $t_{i+1} - t_i = \alpha\theta_i + (1-\alpha)\theta_{i+1} = \theta_{i+1} + \alpha(\theta_i - \theta_{i+1})$. Let $\varepsilon > 0$ be such that $\theta_i > \theta_{i+1} + \varepsilon > \theta_{i+1} + \alpha(\theta_i - \theta_{i+1})$ and $\theta'_{i+1} = \theta_{i+1} + \varepsilon$. Let θ' be the waiting cost vector obtained from θ by replacing θ_{i+1} with θ'_{i+1} and $(\sigma', t') \in \varphi(\theta')$.

By *positive cost monotonicity,* all agents except $i+1$ weakly gain. Since the increase of θ_{i+1} to θ'_{i+1} does not change the efficient queue, that is, for all $i \in N, \sigma'(i) = i$, this is possible only if for all $j \neq i+1, t'_j \geq t_j$. By *budget balance,* $t'_{i+1} \leq t_{i+1}$. Altogether, $t'_{i+1} - t'_i \leq t_{i+1} - t_i = \theta_{i+1} + \alpha(\theta_i - \theta_{i+1}) < \theta_{i+1} + \varepsilon = \theta'_{i+1} < \theta_i$, which contradicts the conclusion of Theorem 5.1.

Case 2-2: For all $i = N \backslash \{n\}, t_{i+1} - t_i = \theta_i$. Let $i = N \backslash \{1, n\}$, and θ'_i be such that $\theta_{i-1} > \theta'_i > \theta_i$. Let θ' be the waiting cost vector obtained from θ by replacing θ_i with θ'_i and $(\sigma', t') \in \varphi(\theta')$. By *positive cost monotonicity,* all agents except i weakly gain. Since the increase of θ_i to θ'_i does not change the efficient queue, this is possible only if for all $j \neq i$, $t'_j \geq t_j$. By *budget balance*, $t'_i \leq t_i$. Altogether, $t'_i - t'_{i-1} \leq t_i - t_{i-1} = \theta_{i-1}$. If $t'_i - t'_{i-1} < \theta_{i-1}$, then we go back to Case 2-1 and obtain the desired conclusion. On the other hand, if $t'_i - t'_{i-1} = \theta'_i = \theta_{i-1}$, by *budget balance,* we deduce that for all $j \in N$, $t'_j = t_j$. In particular, $t'_{i+1} - t'_i = t_{i+1} - t_i = \theta_i$. Since $\theta'_i > t'_{i+1} - t'_i > \theta_{i+1}$, we go back to Case 2-1 and obtain the desired conclusion. □

Remark 5.2 For $|N| = 2$, it is clear from the proof that there is only one rule satisfying *efficiency, Pareto indifference, no-envy,* and *negative cost monotonicity.*[8] It is obtained by setting $t_1 = -\frac{\theta_2}{2}$ and $t_2 = \frac{\theta_2}{2}$, which is the allocation chosen by the minimal transfer rule. The conclusion follows by noting that for $|N| = 2$, the minimal transfer rule satisfies the four axioms. A parallel observation can be made for the maximal transfer rule: for $|N| = 2$, the maximal transfer rule is the only rule satisfying *efficiency, Pareto indifference, no-envy,* and *positive cost monotonicity.*

Next, we ask whether there is a rule satisfying *efficiency* and *no-envy* together with either one of two independence requirements (Sect. 4.2): *independence of preceding costs* requires that an increase in an agent's waiting cost should not affect the agents following her, whereas *independence of following costs* requires that a decrease in an agent's waiting cost should not affect the agents preceding her. Once again, we obtain negative results.

Theorem 5.3 *Let $|N| \geq 3$. Then, there is no rule satisfying efficiency, no-envy, and either independence of preceding costs or independence of following costs.*

Proof

(i) Let φ be a rule satisfying *efficiency, no-envy,* and *independence of preceding costs.* Let $N \in \mathcal{N}$, $\theta \in \mathcal{Q}^N$, and $(\sigma, t) \in \varphi(\theta)$ be such that $N = \{1, \ldots, n\}$ with $n \geq 3$ and $\theta_1 > \theta_2 > \cdots > \theta_n$. By *queue-efficiency,* for all $i \in N$, $\sigma(i) = i$. From Theorem 5.1, for all $i = 1, \ldots, n-1$, $\theta_i \geq t_{i+1} - t_i \geq \theta_{i+1}$.
Now let θ' be such that $\theta'_1 = \theta'_2$ and that for all $i = 3, \ldots, n$, $\theta'_i = \theta_i$. Let $(\sigma', t') \in \varphi(\theta')$. Furthermore, we assume that $\theta'_2 > \max\{\theta_3, 2(t_3 - \theta_3) - (t_1 + t_2)\}$. By *independence of preceding costs,* for all $i = 3, \ldots, n$, $t'_i = t_i$. First, we consider the case $\sigma_1 = 1$ and $\sigma_2 = 2$. By *no-envy,* $t'_2 - t'_1 = \theta'_2$. Therefore, $t'_2 - t'_1 = \theta'_2 > 2(t_3 - \theta_3) - (t_1 + t_2)$. Since *budget balance* requires that $t'_1 + t'_2 = t_1 + t_2$, we have $\theta'_3 = \theta_3 > t_3 - t'_2 = t'_3 - t'_2$, contradicting the conclusion of Theorem 5.1. The case $\sigma_1 = 2$ and $\sigma_2 = 1$ can be handled in a similar way.
(ii) Let φ be a rule satisfying *efficiency, no-envy,* and *independence of following costs.* Let $N \in \mathcal{N}$, $\theta \in \mathcal{Q}^N$, and $(\sigma, t) \in \varphi(\theta)$ be such that $N = \{1, \ldots, n\}$ with $n \geq 3$, $\theta_1 = \theta_2 \geq \theta_3 \geq \cdots \geq \theta_n$ and $\theta_3 > 0$. By *queue-efficiency,* for all $i \in N \backslash \{1, 2\}$, $\sigma(i) = i$. First, we consider the case $\sigma_1 = 1$ and $\sigma_2 = 2$.

[8] If *Pareto indifference* is not imposed, then it is possible to choose only one *efficient* queue when two agents have equal waiting costs.

From Theorem 5.1, for all $i = 1, \ldots, n-1$, $\theta_i \geq t_{i+1} - t_i \geq \theta_{i+1}$. In particular, $t_2 - t_1 = \theta_1$ and $t_n \geq \cdots \geq t_3 > t_2$.

Now let θ' be such that $\theta'_1 = \theta_1$ and $\theta'_2 = \cdots = \theta'_n = 0$. By *independence of following costs*, $t'_1 = t_1$. By *budget balance* and *no-envy*, for all $i = 2, \ldots, n$, $t'_i = \frac{1}{n-1}\sum_{j=2}^{n} t_j$, which implies that $t'_2 > t_2$. Altogether, $t'_2 - t'_1 > t_2 - t_1 = \theta_1 = \theta'_1$, which contradicts the conclusion of Theorem 5.1. The case $\sigma_1 = 2$ and $\sigma_2 = 1$ can be handled in a similar way. □

Remark 5.3 As in Remark 5.2, these impossibility results do not hold if $|N| = 2$. Then, the minimal transfer rule satisfies *efficiency, no-envy,* and *independence of preceding costs,* and the maximal transfer rule satisfies *efficiency, no-envy,* and *independence of following costs.* Moreover, by imposing *Pareto indifference* additionally, the rules can be characterized.

5.4 Adjusted No-Envy and Backward/Forward No-Envy

Given the negative results presented in Sect. 5.3, we propose modifications of *no-envy,* which can be imposed in the context of queueing problems. As we show here, these modifications have a significant effect since we can recover positive results.

To apply *no-envy,* each agent is supposed to reevaluate what any other agents consume. In the queueing problem, an allocation consists of agents' positions in the queue and their transfers, and a rule determines the transfers by agents' positions and waiting costs. If two agents interchange their positions in the queue, then their transfers would not be the same as before because their waiting costs are different. After this adjustment, if an agent does not envy the other agents, we say that the rule satisfies *adjusted no-envy.* To state the requirement formally, we introduce some notation. Given $N \in \mathcal{N}$, $\theta \in \mathcal{Q}^N$, $(\sigma, t) \in \varphi(\theta)$, and $i, j \in N$, let σ^{ij} be the queue obtained from σ by interchanging σ_i and σ_j and t^{ij} be the transfer vector obtained when the rule is applied to σ^{ij}. Since this queue is not efficient in general, strictly speaking, we need to generalize our notion of a rule so that it be applicable to any, not necessarily efficient, queue. For simplicity, we abuse our definition and apply the rule to any queue.

Adjusted no-envy: For all $N \in \mathcal{N}$, all $\theta \in \mathcal{Q}^N$, all $(\sigma, t) \in \varphi(\theta)$, and all $i, j \in N$, $u_i(\sigma_i, t_i; \theta_i) \geq u_i(\sigma^{ij}_i, t^{ij}_i; \theta_i)$.

It is interesting to note that both the minimal and the maximal transfer rules satisfy this requirement.

Proposition 5.2 *The minimal and the maximal transfer rules satisfy adjusted no-envy.*

Proof

(i) We show that the minimal transfer rule satisfies *adjusted no-envy*. Let $N \in \mathcal{N}$, $\theta \in \mathcal{Q}^N$, and $(\sigma^M, t^M) \in \varphi^M(\theta)$ be such that $N = \{1, \ldots, n\}$ and

$\theta_1 \geq \theta_2 \geq \cdots \geq \theta_n$. To simplify notation, for all $i \in N$, we set $\sigma_i^M = i$. Also, we do not attach the superscript M to σ and t. Let $i, j \in N$. We may assume, without loss of generality, that $j = i + k$ for some $k \in \mathbf{N}$.

First, we show that i does not envy $j = i + k$ if the adjustment is made. Note that

$$u_i(\sigma_i, t_i; \theta_i) = (i-1)\frac{\theta_i}{2} - \sum_{\ell=i+1}^{n} \frac{\theta_\ell}{2} - (i-1)\theta_i,$$

and

$$u_i(\sigma_i^{ij}, t_i^{ij}; \theta_i) = (i+k-1)\frac{\theta_i}{2} - \sum_{\ell=i+k+1}^{n} \frac{\theta_\ell}{2} - (i+k-1)\theta_i.$$

Therefore,

$$u_i(\sigma_i, t_i; \theta_i) - u_i(\sigma_i^{ij}, t_i^{ij}; \theta_i) = k\frac{\theta_i}{2} - \sum_{\ell=i+1}^{i+k} \frac{\theta_\ell}{2} \geq 0,$$

as desired. Similarly, we can show that $j = i+k$ does not envy i if the adjustment is made.

(ii) Now we show that the maximal transfer rule satisfies *adjusted no-envy*. Let $N \in \mathcal{N}$, $\theta \in \mathcal{Q}^N$, and $(\sigma^C, t^C) \in \varphi^C(\theta)$ be such that $N = \{1, \ldots, n\}$ and $\theta_1 \geq \theta_2 \geq \cdots \geq \theta_n$. To simplify notation, for all $i \in N$, we set $\sigma_i^C = i$. Also, we do not attach the superscript C to σ and t. Let $i, j \in N$. We may assume, without loss of generality, that $j = i + k$ for some $k \in \mathbf{N}$.

First, we show that i does not envy $j = i + k$ if the adjustment is made. Note that

$$u_i(\sigma_i, t_i; \theta_i) = \sum_{\ell=1}^{i-1} \frac{\theta_\ell}{2} - (n-i)\frac{\theta_i}{2} - (i-1)\theta_i,$$

and

$$u_i(\sigma_i^{ij}, t_i^{ij}; \theta_i) = \sum_{\ell=1}^{i+k-1} \frac{\theta_\ell}{2} - \frac{\theta_i}{2} + \frac{\theta_j}{2} - (n-i-k)\frac{\theta_i}{2} - (i+k-1)\theta_i.$$

Therefore,

$$u_i(\sigma_i, t_i; \theta_i) - u_i(\sigma_i^{ij}, t_i^{ij}; \theta_i) = -\frac{\theta_j}{2} - \sum_{\ell=i+1}^{i+k-1} \frac{\theta_\ell}{2} + k\frac{\theta_i}{2} \geq 0,$$

as desired. Similarly, we can show that $j = i+k$ does not envy i if the adjustment is made. □

Now we propose two weakenings of *no-envy*, which require that an agent should not envy other agents at least in one direction. *Backward no-envy* requires that an agent should not envy the agents with lower waiting costs (therefore, following her in the efficient queue with the possible exception of agents with the same waiting cost), whereas *forward no-envy* requires that an agent should not envy the agents with higher waiting costs (therefore, preceding her in the efficient queue with the possible exception of agents with the same waiting cost). In the language of Theorem 5.1, *backward no-envy,* together with *efficiency,* requires that for all $\sigma_i = 1, \ldots, n-1$, $\theta_{\sigma_i} \geq t_{\sigma_i+1} - t_{\sigma_i}$. On the other hand, *forward no-envy,* together with *efficiency,* requires that $t_{\sigma_i+1} - t_{\sigma_i} \geq \theta_{\sigma_i+1}$. The minimal transfer rule satisfies *backward no-envy,* while the maximal transfer rule satisfies *forward no-envy.*

Backward no-envy: For all $N \in \mathcal{N}$, all $\theta \in \mathcal{Q}^N$, all $(\sigma, t) \in \varphi(\theta)$, and all i, $j \in N$, if $\theta_i \geq \theta_j$, then $u_i(\sigma_i, t_i; \theta_i) \geq u_j(\sigma_j, t_j; \theta_i)$.

Forward no-envy: For all $N \in \mathcal{N}$, all $\theta \in \mathcal{Q}^N$, all $(\sigma, t) \in \varphi(\theta)$, and all $i, j \in N$, if $\theta_i \leq \theta_j$, then $u_i(\sigma_i, t_i; \theta_i) \geq u_j(\sigma_j, t_j; \theta_i)$.

Proposition 5.3 *The minimal transfer rule satisfies backward no-envy and the maximal transfer rule satisfies forward no-envy.*

Proof Let $N \in \mathcal{N}$, $\theta \in \mathcal{Q}^N$, and $(\sigma, t) \in \varphi(\theta)$ be such that $N = \{1, \ldots, n\}$ and $\theta_1 \geq \theta_2 \geq \cdots \geq \theta_n$. First, we show that the minimal transfer rule satisfies *backward no-envy*. To simplify notation, for all $i \in N$, we set $\sigma_i^M = i$. Also, we do not attach the superscript M to σ and t. Since the minimal transfer rule satisfies *efficiency* and *equal treatment of equals,* it is enough to show that for $i = 1, \ldots, n-1$, if $\theta_i > \theta_{i+1}$, i does not envy $i+1$. From the definition of the minimal transfer rule,

$$u_i(\sigma_i, t_i; \theta_i) = -(\sigma_i - 1)\theta_i + t_i = -(i-1)\theta_i + (i-1)\frac{\theta_i}{2} - \sum_{j=i+1}^{n} \frac{\theta_j}{2},$$

and

$$u_i(\sigma_{i+1}, t_{i+1}; \theta_i) = -(\sigma_{i+1} - 1)\theta_i + t_{i+1} = -i\theta_i + i\frac{\theta_{i+1}}{2} - \sum_{j=i+2}^{n} \frac{\theta_j}{2}.$$

Therefore,

$$u_i(\sigma_i, t_i; \theta_i) - u_i(\sigma_{i+1}, t_{i+1}; \theta_i) = (i+1)(\frac{\theta_i}{2} - \frac{\theta_{i+1}}{2}) \geq 0,$$

as desired.

Next, we show that the maximal transfer rule satisfies *forward no-envy*. To simplify notation, for all $i \in N$, we set $\sigma_i^C = i$. Also, we do not attach the superscript

C to σ and t. Since the maximal transfer rule satisfies *efficiency* and *equal treatment of equals,* it is enough to show that for $i = 1, \ldots, n-1$, if $\theta_{i-1} > \theta_i$, i does not envy $i-1$. From the definition of the maximal transfer rule,

$$u_i(\sigma_i, t_i; \theta_i) = -(\sigma_i - 1)\theta_i + t_i = -(i-1)\theta_i + \sum_{j=1}^{i-1} \frac{\theta_j}{2} - (n - \sigma_i)\frac{\theta_i}{2},$$

and

$$u_i(\sigma_{i-1}, t_{i-1}; \theta_i) = -(\sigma_{i-1} - 1)\theta_i + t_{i-1} = -(i-2)\theta_i + \sum_{j=1}^{i-2} \frac{\theta_j}{2} - (n - \sigma_{i-1})\frac{\theta_{i-1}}{2}.$$

Therefore,

$$u_i(\sigma_i, t_i; \theta_i) - u_i(\sigma_{i-1}, t_{i-1}; \theta_i) = (n - i + 2)(\frac{\theta_{i-1}}{2} - \frac{\theta_i}{2}) \geq 0,$$

as desired. □

5.5 Other Fairness Requirements

Although no-envy plays an important role in the literature on the fairness, there are other interesting concepts. The main ones are the *identical preferences lower bound* and *egalitarian equivalence*. The *identical preferences lower bound* (Moulin 1990) requires that each agent should be at least as well off as she would be, under *efficiency* and *equal treatment of equals,* if all other agents had the same preferences. *Egalitarian equivalence* (Pazner and Schmeidler 1978) requires that there should be a reference bundle such that each agent enjoys the same utility between her bundle and that reference bundle. Now we formally introduce these axioms. Given $N \in \mathcal{N}$ and $\theta \in \mathcal{Q}^N$, an allocation $(\sigma, t) \in Z(\theta)$ satisfies the *identical preferences lower bound* if for all $i \in N$, $u(\sigma_i, t_i; \theta_i) \geq -\frac{|N|-1}{2}\theta_i$. It is *egalitarian equivalent* if there is a reference bundle (σ_0, t_0) such that for all $i \in N$, $u_i(\sigma_i, t_i; \theta_i) = u_i(\sigma_0, t_0; \theta_i)$. Let $B_{id}(\theta)$ be the set of all allocations meeting the *identical preferences lower bound* and $EE(\theta)$ be the set of all egalitarian equivalent allocations for $\theta \in \mathcal{Q}^N$.

Identical preferences lower bound: For all $N \in \mathcal{N}$, all $\theta \in \mathcal{Q}^N$, and all $(\sigma, t) \in \varphi(\theta)$, $(\sigma, t) \in B_{id}(\theta)$.

Egalitarian equivalence: For all $N \in \mathcal{N}$, all $\theta \in \mathcal{Q}^N$, and all $(\sigma, t) \in \varphi(\theta)$, $(\sigma, t) \in EE(\theta)$.

Remark 5.4 In economies with indivisible goods,[9] when there are as many objects as agents, *budget balance* and *no-envy* together imply the *identical preferences lower bound* (Bevia 1996). Moreover, if there are only two agents, then *budget balance* and *identical preferences lower bound* together imply *no-envy.* A similar observation can be made for queueing problems.

Now we investigate whether a rule can satisfy *no-envy* and *egalitarian equivalence* together. If there are only two agents, then any rule satisfying *efficiency* and *egalitarian equivalence* satisfies *no-envy*. Moreover, if there are only three agents, then by choosing the middle position as a part of the reference bundle, we can establish the existence of a rule satisfying *efficiency, no-envy,* and *egalitarian equivalence*. However, the positive result does not generalize to problems with more than three agents.

Proposition 5.4 (Chun et al. 2014) *Let* $|N| \geq 4$*. Then, there is no rule satisfying no-envy and egalitarian equivalence together.*

Proof Let φ be a rule satisfying *no-envy* and *egalitarian equivalence*. Let $N \in \mathcal{N}$ and $\theta \in \mathcal{Q}^N$ be such that $N = \{1, \ldots, n\}$ and $\theta_1 > \cdots > \theta_n$. By *egalitarian equivalence,* there exists (σ_0, t_0) such that for all $i \in N$, $-(\sigma_i - 1)\theta_i + t_i = -(\sigma_0 - 1)\theta_i + t_0$. Rewriting this, we have for all $i \in N$,

$$t_i = (\sigma_i - \sigma_0)\theta_i + t_0. \tag{5.2}$$

Choose two agents i and $i+1$. As shown in Proposition 5.1, *no-envy* implies *queue-efficiency*, and hence, $\sigma_{i+1} = \sigma_i + 1$. From Eq. (5.1), it follows that $\theta_{i+1} \leq t_{i+1} - t_i \leq \theta_i$. From Eq. (5.2), $\theta_{i+1} \leq (\sigma_{i+1} - \sigma_0)\theta_{i+1} - (\sigma_i - \sigma_0)\theta_i \leq \theta_i$, which implies that

$$0 \leq (\sigma_0 - \sigma_i)(\theta_i - \theta_{i+1}) \leq \theta_i - \theta_{i+1}. \tag{5.3}$$

Since $\theta_i > \theta_{i+1}$, it follows that $0 \leq \sigma_0 - \sigma_i \leq 1$. Note that the selection of i and $i+1$ has been arbitrary. By choosing $i = 1$, we obtain $0 \leq \sigma_0 - 1 \leq 1$ which implies that $\sigma_0 \in \{1, 2\}$. By choosing $i = n - 1$, we obtain $\sigma_0 \in \{n - 1, n\}$. The two restrictions on σ_0 are incompatible when $n \geq 4$. □

In economies with indivisible goods, there is no rule satisfying *object-efficiency, egalitarian equivalence,* and the *identical preferences lower bound* (Thomson 2003). However, in queueing problems, we can construct a rule satisfying *efficiency, egalitarian equivalence,* and the *identical preferences lower bound.*

Proposition 5.5 *If there is an odd number of agents, then there is at least one efficient and egalitarian equivalent allocation meeting the identical preferences lower bound. If there is an even number of agents, then there are at least two*

[9] Note that *budget balance* is imposed as a part of the feasibility requirement in Bevia (1996).

efficient and egalitarian equivalent allocations meeting the identical preferences lower bound.

Proof Let $N \equiv \{1, \ldots, n\}$ be such that n is an odd number and $\theta \equiv (\theta_i)_{i \in N}$ be such that $\theta_1 \geq \theta_2 \geq \cdots \geq \theta_n$. To simplify notation, for all $i \in N$, we set $\sigma_i = i$. Let $i \equiv \frac{n+1}{2}$ and $z_i \equiv (\sigma_i, t_i)$. We will determine the value of t_i later after considering the budget constraint.

For each $j \in N$, let $z_j \equiv (\sigma_j, t_i + (j - i)\theta_j)$. Now we calculate t_i by solving $nt_i + \sum_{j \in N}(j - i)\theta_j = 0$, which gives $t_i = -\frac{1}{n}\sum_{j \in N}(j - i)\theta_j$. Since $\theta_1 \geq \theta_2 \geq \cdots \geq \theta_n$, then $t_i \geq 0$. It is obvious that $z \equiv (z_j)_{j \in N}$ is efficient. Since for all $j \in N$, $u_j(z_j; \theta_j) = u_j(z_i; \theta_j)$, it is also *egalitarian equivalent.*

To prove that z satisfies the *identical preferences lower bound,* we need to show that for each $j \in N$, $u(z_j; \theta_j) \geq -\frac{n-1}{2}\theta_j$. For each $j \in N$, $u_j(z_j; \theta_j) = -(j - 1)\theta_j + t_i + (j - i)\theta_j \geq -(i - 1)\theta_j = -(\frac{n+1}{2} - 1)\theta_j = -\frac{n-1}{2}\theta_j$, as desired.

On the other hand, if n is an even number, then we choose either $\frac{n}{2}$ or $\frac{n}{2} + 1$ as reference positions. If $i = \frac{n}{2}$, we can show that $t_i \geq -\frac{1}{2}\theta_n$. Therefore, for each $i \in N$, $u_j(z_j; \theta_j) = -(j-1)\theta_j + t_i + (j-i)\theta_j \geq -(i-1)\theta_j - \frac{1}{2}\theta_n \geq -(i-1)\theta_j - \frac{1}{2}\theta_j = -(\frac{n}{2} - 1 + \frac{1}{2})\theta_j = -\frac{n-1}{2}\theta_j$, as desired. If $n = \frac{n}{2} + 1$, we can show that for each $j \in N$, $t_i \geq \frac{1}{2}\theta_1 \geq \frac{1}{2}\theta_j$. From a similar calculation, we obtain the desired conclusion. □

5.6 Concluding Remarks

By investigating the implications of *no-envy* in queueing problems, we establish various results. Our main negative results are there is no rule satisfying *efficiency, no-envy,* and either one of two *cost monotonicities.* These results should be compared with the impossibility result in Moulin and Thomson (1988): in the classical economies, there is no rule satisfying *Pareto optimality,*[10] *no-envy,* and *resource monotonicity.*[11] Since our problem is very different from theirs, there is no direct logical implications between two results. However, at least conceptually, we are faced with the same difficulties: axioms of *efficiency, no-envy,* and *monotonicity* are not compatible.[12]

To remedy this situation, modifications of *no-envy* are proposed as fairness requirements in queueing problems. Although the implications of *backward* and *forward no-envy* are clear, it is an open question what the implications of *adjusted no-envy* in queueing problems are. In particular, its relation to *no-envy* needs to be analyzed.

[10] *Pareto optimality* requires that there is no feasible allocation which makes every agent better off and at least one agent strictly better off.

[11] *Resource monotonicity* requires that an increase in resources should not hurt any agent.

[12] For a possibility result, see Alkan et al. (1991).

References

Alkan, A., Demange, G., & Gale. D. (1991). Fair allocation of indivisible goods and criteria of justice. *Econometrica, 59*, 1023–1039.
Bevia, C. (1996). Identical preferences lower bound solution and consistency in economies with indivisible goods. *Social Choice and Welfare, 13*, 113–126.
Chun, Y. (2005). Characterizations of the symmetric Groves rule in queueing problems. Mimeo.
Chun, Y. (2006). No-envy in queueing problems. *Economic Theory, 29*, 151–162.
Chun, Y., Mitra, M., & Mutuswami, S. (2014). Egalitarian equivalence and strategyproofness in the queueing problem. *Economic Theory, 56*, 425–442.
Foley, D. (1967). Resource allocation and the public sector. *Yale Economic Essays, 7*, 45–98.
Moulin, H. (1990). Fair division under joint ownership: Recent results and open problems. *Social Choice and Welfare, 7*, 149–170.
Moulin, H., & Thomson, W. (1988). Can everyone benefit from growth? Two difficulties. *Journal of Mathematical Economics, 17*, 339–345.
Pazner, E., & Schmeidler, D. (1978). Egalitarian equivalent allocations: A new concept of economic equity. *Quarterly Journal of Economics, 92*, 671–687.
Svensson, L. G. (1983). Large indivisibles: An analysis with respect to price equilibrium and fairness. *Econometrica, 51*, 939–954.
Thomson, W. (2003). On monotonicity in economies with indivisible goods. *Social Choice and Welfare, 21*, 195–205.
Thomson, W. (2005). The theory of fair allocation. University of Rochester, Book Manuscript.
Thomson, W., & Varian, H. (1985). Theories of justice based on symmetry, Chapter 4. In L. Hurwicz, D. Schmeidler, & H. Sonnenschein (Eds.), *Social goals and social organization: essays in memory of E. Pazner* (pp. 107–129). Cambridge: Cambridge University Press.

Chapter 6
Strategyproofness

6.1 Introduction

Strategyproofness requires that an agent should not have an incentive to misrepresent her waiting cost no matter what she believes other agents to be doing. Therefore, truthful reporting of the waiting cost is a weakly dominant strategy for all agents. This condition has been studied extensively in the literature.[1] In this chapter, we investigate its implications in the context of queueing problems.

The classic result of Holmström (1979) implies in the context of queueing problems that a rule satisfies *queue-efficiency* and *strategyproofness* if and only if it is a VCG rule.[2] *Queue-efficiency* requires that the selected queue should minimize the aggregate waiting cost. Imposing additional axioms gives us a subfamily of rules from the class of VCG rules.

By additionally imposing *equal treatment of equals,* which requires that two agents with the same waiting costs should end up with the same utilities, we characterize anonymous members of the VCG rules. This subfamily includes the symmetrically balanced VCG rule (Kayi and Ramaekers 2010; Mitra 2001; Suijs 1996) and the pivotal and the reward-based pivotal rules (Mitra and Mutuswami 2011). The symmetrically balanced VCG rule can be singled out from the family by requiring a rule to be *budget balanced.* On the other hand, the pivotal and the reward-based pivotal rules can be characterized by imposing independence axioms introduced by Maniquet (2003) and Chun (2006a) (see Sect. 4.2 for more explanation on these axioms). Finally, by generalizing the independence axioms, we can characterize the entire class of *k*-pivotal rules.

[1]The literature on strategyproofness is too large to give a comprehensive list of references. A recent review of this literature, along with a list of references, can be found in Barberà (2011) and Thomson (2013).

[2]The family of VCG rules is due to Vickrey (1961), Clarke (1971), and Groves (1973).

© Springer International Publishing Switzerland 2016
Y. Chun, *Fair Queueing*, Studies in Choice and Welfare,
DOI 10.1007/978-3-319-33771-5_6

This chapter is organized as follows.[3] In Sect. 6.2, we characterize anonymous VCG rules by additionally imposing *equal treatment of equals*. From this subfamily of VCG rules, we single out the symmetrically balanced VCG rule, the pivotal and the reward-based pivotal rules, and then the entire class of k-pivotal rules in Sect. 6.3. Concluding remarks are given in Sect. 6.4.

6.2 Strategyproofness and the VCG Rules

To use Holmström's result, we need to assume a rule to be single-valued. Note that so far a rule is assumed to be multi-valued. To distinguish between a multi-valued rule and a single-valued rule, we continue to use φ for a multi-valued rule and newly introduce μ to denote a single-valued rule.

A (single-valued) *rule* is a function μ which associates to each problem θ, a tuple $\mu(\theta) = (\sigma, t) \in \Sigma(N) \times \mathbf{R}^n$ where σ is the selected queue and $t = (t_i)_{i \in N}$ is the vector of transfers. In this chapter, we fix the set of agents and change the profile of waiting costs. To indicate the dependence on the profile θ, we denote the allocation as $\mu(\theta) \equiv (\sigma(\theta), t(\theta))$.[4] For each agent $i \in N$, let $\mu_i(\theta) = (\sigma_i(\theta), t_i(\theta))$ be agent i's allocation for the problem θ and $u_i(\mu_i(\theta); \theta'_i) = -(\sigma_i(\theta) - 1)\theta'_i + t_i(\theta)$ be agent i's utility when the profile of announced waiting costs is θ and her true waiting cost is θ'_i.

Now we introduce two axioms which play an important role in this chapter. Our first axiom is *queue-efficiency,* which requires that a rule should choose an efficient queue for all problems (see Sect. 2.3 for definition). Since we assume a rule to be single-valued, *queue-efficiency* can be rewritten as follows.

Queue-efficiency: For all $N \in \mathcal{N}$ and all $\theta \in \mathcal{Q}^N$, $\sigma(\theta) \in Eff(\theta)$.

Remark 6.1 The efficient queue is unique at all profiles where no two agents have identical waiting costs. However, if there are some agents with the same waiting cost, then the efficient queue is not unique. Since *queue-efficiency* is the only axiom that we impose on the queue, it is not clear which queue we should choose for the profile. In Chaps. 6 and 7, we implicitly assume the existence of a tie-breaking rule, which selects an efficient queue whenever there is more than one such queue. We assume that there is an order of the agents which is used to break ties. The same order is used to break ties when a queue involving subsets of agents has to be selected. Let $\mathcal{T}$ be the set of all possible tie-breaking rules for N and τ be a typical element of $\mathcal{T}$. We note that our result applies for any choice of tie-breaking rule.

---

[3]This chapter is based mainly on Chun et al. (2011, 2014). Excerpts from Chun et al. (2014) are reprinted with kind permission of Elsevier.

[4]Since t depends on the choice of the queue, we should denote the transfers by $t(\sigma(\theta))$ instead of $t(\theta)$. Note that our (single-valued) rule chooses a unique queue which in turn determines the unique transfers. Therefore, we abuse the notation and write $t(\theta)$.

Our second axiom is *strategyproofness,* which requires that an agent cannot strictly gain by misrepresenting her waiting cost no matter what she believes other agents to be doing.

Strategyproofness: For all $N \in \mathcal{N}$, all $\theta, \theta' \in \mathcal{Q}^N$, all $i \in N$, all $(\sigma, t) \in \varphi(\theta)$, and $(\sigma', t') \in \varphi(\theta')$, if for all $j \in N \setminus \{i\}$, $\theta_j = \theta'_j$, then $u_i(\sigma_i, t_i; \theta_i) \geq u_i(\sigma'_i, t'_i; \theta_i)$.

Once again, since we assume the single-valuedness of a rule, *strategyproofness* can be rewritten in the following simple way.

Strategyproofness: For all $N \in \mathcal{N}$, all $\theta \in \mathcal{Q}^N$, all $i \in N$, and all $\theta'_i \in \mathbf{R}_+$, $u_i(\mu_i(\theta); \theta_i) \geq u_i(\mu_i(\theta'_i, \theta_{N\setminus\{i\}}); \theta_i)$.

Remark 6.2 Holmström (1979) shows that when preferences are quasi-linear and the domain of types is convex, the VCG rules are the only ones satisfying *queue-efficiency* and *strategyproofness*. For queueing problems, the preferences are completely specified by the profile of waiting costs. Since this is $\mathbf{R}^n_+$, it follows that a rule satisfies *queue-efficiency* and *strategyproofness* if and only if it is a VCG rule.

We use the following notation. For all $N \in \mathcal{N}$ and all $\theta \in \mathcal{Q}^N$, suppose there is an initial queue $\sigma(\theta)$ and agent $i \in N$ leaves the queue. We define the "induced" queue $\sigma(\theta_{N\setminus\{i\}})$ (of length $n-1$) for the agents in $N \setminus \{i\}$ as follows:

$$\sigma_j(\theta_{N\setminus\{i\}}) = \begin{cases} \sigma_j(\theta) & \text{if } j \in P_i(\theta), \\ \sigma_j(\theta) - 1 & \text{if } j \in F_i(\theta). \end{cases} \tag{6.1}$$

In words, $\sigma(\theta_{N\setminus\{i\}})$ is the queue formed by removing agent i and moving all agents behind her up by one position. It is easy to see that $\sigma(\theta_{N\setminus\{i\}})$ is efficient in $N \setminus \{i\}$ for the profile $\theta_{N\setminus\{i\}}$ if $\sigma(\theta)$ is efficient for the profile θ.

We now formally define the VCG rules.

VCG rule associated with g_i, μ^{g_i}: For all $N \in \mathcal{N}$ and all $\theta \in \mathcal{Q}^N$, $\mu^{g_i}(\theta) = (\sigma(\theta), t(\theta))$ is defined as $\sigma(\theta) \in \mathit{Eff}(\theta)$, and for all $i \in N$,

$$t_i(\theta) = -\sum_{j \in F_i(\sigma)} \theta_j + g_i(\theta_{N\setminus\{i\}}). \tag{6.2}$$

Remark 6.3 The standard way of specifying the VCG transfers is

$$\forall \theta, \forall i \in N : t_i(\theta) = -\sum_{j \neq i} (\sigma_j(\theta) - 1)\theta_j + h_i(\theta_{N\setminus\{i\}}).$$

Since we can write without loss of generality that

$$\forall \theta, \forall i \in N : h_i(\theta_{N\setminus\{i\}}) = \sum_{j \neq i} (\sigma_j(\theta_{N\setminus\{i\}}) - 1)\theta_j + g_i(\theta_{N\setminus\{i\}}),$$

we have

$$t_i(\theta) = -\sum_{j\neq i}(\sigma_j(\theta) - 1)\theta_j + h_i(\theta_{N\setminus\{i\}})$$
$$= -\sum_{j\neq i}(\sigma_j(\theta) - 1)\theta_j + \sum_{j\neq i}(\sigma_j(\theta_{N\setminus\{i\}}) - 1)\theta_j + g_i(\theta_{N\setminus\{i\}})$$
$$= -\sum_{j\in F_i(\sigma(\theta))}\theta_j + g_i(\theta_{N\setminus\{i\}}),$$

the desired expression.

Remark 6.4 For a VCG rule, an agent's utility is independent of the tie-breaking rule. Suppose that $\sigma, \sigma' \in \mathit{Eff}(\theta)$ with $\sigma_i \neq \sigma'_i$. By *queue-efficiency,* all agents whose waiting cost is θ_i occupy the same set of consecutive queue positions in both queues. Therefore, by Eq. (6.2), the utility of agent i in the two queues can differ only through differences in $g_i(\theta_{N\setminus\{i\}})$ across tie-breaking rules, but since g_i is independent of θ_i, it cannot depend on the tie-breaking rule for θ either.

Equal treatment of equals requires that two agents with the same waiting cost should end up with the same utilities. We investigate the implications of imposing *equal treatment of equals* together with *queue-efficiency* and *strategyproofness* and characterize the family of anonymous VCG rules.

Anonymous VCG rule associated with g, μ^g: For all $N \in \mathcal{N}$ and all $\theta \in \mathcal{Q}^N$,

(1) μ^g is a VCG rule.
(2) For all $i \in N$, g_i is symmetric, i.e., $g_i(x) = g_i(y)$ whenever x and y are permutations of one another.
(3) For all $i, j \in N$ such that $\theta_i = \theta_j$, $g_i(\theta_{N\setminus\{i\}}) = g_j(\theta_{N\setminus\{j\}})$.

Remark 6.5 Given (2) and (3), we can write $g_i = g$ for all $i \in N$.

Our characterization result follows.

Proposition 6.1 *A rule satisfies queue-efficiency, equal treatment of equals, and strategyproofness if and only if it is an anonymous VCG rule.*

Proof It is easy to show that an anonymous VCG rule satisfies *queue-efficiency, equal treatment of equals,* and *strategyproofness*. Conversely, let μ be a rule satisfying the three axioms. From Remark 6.2, if a rule satisfies *queue-efficiency* and *strategyproofness,* then it is a VCG rule. Hence, the transfers are given by

$$\forall\theta, \forall i \in N: \quad t_i(\theta) = -\sum_{j\in F_i(\sigma(\theta))}\theta_j + g_i(\theta_{N\setminus\{i\}}).$$

Let $\theta_{N\setminus\{i\}} \in \mathbf{R}_+^{n-1}$. Consider $(\theta_j, \theta_{N\setminus\{i\}}) \in \mathbf{R}_+^n$ where i and j announce θ_j and the others announce $\theta_{N\setminus\{i,j\}}$. Using *equal treatment of equals* between i and j, it follows that[5]

$$g_i(\theta_{N\setminus\{i\}}) = g_j(\theta_j, \theta_{N\setminus\{i,j\}}). \tag{6.3}$$

Next, consider $(\theta_j, \theta_k, \theta_{N\setminus\{i,j\}})$ where i announces θ_j, j and k announce θ_k and the others announce $\theta_{N\setminus\{i,j,k\}}$. By *equal treatment of equals,* it follows that

$$g_j(\theta_j, \theta_{N\setminus\{i,j\}}) = g_k(\theta_j, \theta_k, \theta_{N\setminus\{i,j,k\}}). \tag{6.4}$$

It follows from Eqs. (6.3) and (6.4) that

$$g_k(\theta_j, \theta_k, \theta_{N\setminus\{i,j,k\}}) = g_i(\theta_{N\setminus\{i\}}). \tag{6.5}$$

Now, consider $(\theta_k, \theta_{N\setminus\{i\}})$ where i and k announce θ_k and the rest announce $\theta_{N\setminus\{i,k\}}$. Using *equal treatment of equals* between i and k, we obtain

$$g_i(\theta_{N\setminus\{i\}}) = g_k(\theta_k, \theta_{N\setminus\{i,k\}}). \tag{6.6}$$

It follows from Eqs. (6.5) and (6.6) that

$$g_k(\theta_j, \theta_k, \theta_{N\setminus\{i,j,k\}}) = g_k(\theta_k, \theta_j, \theta_{N\setminus\{i,j,k\}}).$$

Observe that the choice of i, j, and k is arbitrary. Since g_k does not depend on θ_k, the above equation effectively shows that a permutation of the announcements of two agents in $N \setminus \{k\}$ while leaving others unchanged does not change g_k. Since we can get from a profile $\theta_{N\setminus\{k\}}$ to an arbitrary permutation through a sequence of pairwise permutations, this proves that g_i, $i = 1, \ldots, n$, is symmetric.

Finally, *equal treatment of equals* implies that for all $i, j \in N$, $i \neq j$, if $\theta_i = \theta_j$, then $g_i(\theta_j, \theta_{N\setminus\{i,j\}}) = g_j(\theta_i, \theta_{N\setminus\{i,j\}})$. This along with symmetry of g_i implies that all the g_i functions are the same, so that we can put for all $i \in N$, $g_i = g$. □

Remark 6.6 In the context of a related but different model of allocating heterogeneous goods, Pápai (2003) (see Observation 3, p. 376 of her paper) shows that the family of *anonymous VCG rules* can be characterized using the property of *no-envy* (Foley 1967) which requires that no agent should end up with a higher utility by consuming what any other agent consumes (see Chap. 5 for its implications in the context of queueing problems).

[5]This step is essentially a minor modification of the argument in Remark 6.4. In particular, by *queue-efficiency,* all agents occupying queue positions between σ_i and σ_j have waiting cost θ_j and by *equal treatment of equals,* they receive the same utilities implying Eq. (6.3).

Remark 6.7 All the axioms are necessary for the result to hold.

(i) Dropping *queue-efficiency*: Let $N = \{1, 2\}$. Let μ^* be a rule such that for all $\theta \in \mathscr{Q}^N$, $\sigma_i(\theta) = i$ and the transfers are given by

$$t_i(\theta) = \begin{cases} -\theta_2/2 \text{ if } i = 1, \\ \theta_1/2 \text{ if } i = 2. \end{cases}$$

It is easy to check this rule satisfies *equal treatment of equals* and *strategyproofness*.

(ii) Dropping *equal treatment of equals*: The entire class of VCG rules become admissible. *Equal treatment of equals* will be violated by choosing the g_i in an asymmetric manner. For example, with $N = \{1, 2\}$, we can choose $g_1(\theta_2) = \theta_2$ and $g_2(\theta_1) = 2\theta_1$.

(iii) Dropping *strategyproofness*: The minimal and the maximal transfer rules (Chun 2006a; Maniquet 2003) satisfy *queue-efficiency* and *equal treatment of equals*.

Remark 6.8 *Anonymity in welfare* requires that a permutation of waiting costs implies a permutation of welfare also. Hashimoto and Saitoh (2012) show that *anonymity in welfare* and *strategyproofness* together imply *queue-efficiency*. It is natural to ask whether *anonymity in welfare* can be weakened to *equal treatment of equals*. However, the rule μ^* in Remark 6.7 (i) shows that *equal treatment of equals* and *strategyproofness* together do not imply *queue-efficiency*.

6.3 Further Characterization Results

We discuss how to single out some interesting rules from the class of anonymous VCG rules characterized in Proposition 6.1.

6.3.1 *The Symmetrically Balanced VCG Rule*

The symmetrically balanced VCG rule defined in Sect. 2.4 is the VCG rule satisfying *budget balance,* which requires that there is no net transfers into or out of the problem. Since we assume in this chapter that a rule is single-valued, we need to modify the definition of the symmetrically balanced VCG rule accordingly.

Symmetrically balanced VCG rule, μ^B: For all $N \in \mathcal{N}$ with $|N| \geq 3$ and all $\theta \in \mathcal{Q}^N$, $\mu^B(\theta) = (\sigma^B(\theta), t^B(\theta))$ is defined as $\sigma^B(\theta) \in Eff(\theta)$, and for all $i \in N$,

$$t_i^B(\theta) = \sum_{j \in P_i(\sigma^B(\theta))} \left(\frac{\sigma_j^B(\theta) - 1}{|N| - 2} \right) \theta_j - \sum_{k \in F_i(\sigma^B(\theta))} \left(\frac{|N| - \sigma_k^B(\theta)}{|N| - 2} \right) \theta_k.$$

This rule is obtained by setting for all $i \in N$,

$$g(\theta_{N \setminus \{i\}}) = \sum_{j \in N \setminus \{i\}} \left(\frac{\sigma_j^B(\theta_{N \setminus \{i\}}) - 1}{n - 2} \right) \theta_j. \tag{6.7}$$

The symmetrically balanced VCG rule has many nice properties. It is *queue-efficient, strategyproof* and *budget balanced,* and hence "first-best" implementable (Mitra 2001; Suijs 1996). It also satisfies *no-envy* (Chun 2006b). Recently, Kayi and Ramaekers (2010, in press) characterize the symmetrically balanced VCG rule by imposing the axioms of *queue-efficiency, Pareto indifference, equal treatment of equals, strategyproofness,* and *budget balance*. *Pareto indifference* requires that for all profiles, if an allocation is chosen by a rule and there is another feasible allocation which gives the same utility to each agent, then this allocation should be chosen by the rule. Chun et al. (in press) provide an alternative simple proof for the characterization. Since we assume the single-valuedness of a rule in this chapter, we can characterize the symmetrically balanced VCG rule without imposing *Pareto indifference*.

Theorem 6.1 *Let $n \geq 3$. A rule satisfies queue-efficiency, equal treatment of equals, strategyproofness, and budget balance if and only if it is the symmetrically balanced VCG rule.*

Proof It is obvious that the symmetrically balanced VCG rule satisfies *queue-efficiency, equal treatment of equals, strategyproofness,* and *budget balance*. To prove the converse statement, let μ be a rule satisfying the four axioms. Let $N \in \mathcal{N}$ be such that $|N| \geq 3$, $\theta' = (\theta'_i)_{i \in N}$ be such that $\theta'_1 > \cdots > \theta'_n$, and $\mu(\theta') = (\sigma(\theta'), t(\theta'))$. We show that

$$g(\theta'_{-n}) = \sum_{i=1}^{n-1} \left(\frac{i-1}{n-2} \right) \theta'_i. \tag{6.8}$$

By *queue-efficiency,* for all $i \in N$, $\sigma_i(\theta') = i$. By Proposition 6.1 and *budget balance,*

$$\sum_{i \in N} g(\theta'_{N \setminus \{i\}}) = \sum_{i \in N} (i - 1)\theta'_i.$$

Let $\Theta = \{\theta'_1, \cdots, \theta'_n\}$. For $k = 1, \ldots, n$, let Θ^k be the set of profiles θ such that

(1) $\theta_i \in \Theta$ for all $i \in N$,
(2) $i < j < k \Longrightarrow \theta_i > \theta_j > \theta'_n$,
(3) $\theta_i = \theta'_n$ if $i \geq k$.

We will prove the following hypothesis by induction. For all $k = 1, \ldots, n$, and all $\theta \in \Theta^k$,

$$g(\theta_{-n}) = \sum_{r=1}^{k-1} \left(\frac{r-1}{n-2}\right) \theta_r + \frac{(n+k-3)(n-k)}{2(n-2)} \theta'_n. \tag{6.9}$$

Step 1: When $k = 1$, the set Θ^1 is a singleton and $\theta = (\theta'_n, \ldots, \theta'_n)$. By *budget balance,*

$$\sum_{i \in N} g(\theta_{-i}) = \frac{n(n-1)}{2} \theta'_n.$$

Since g is symmetric, we can write the above as

$$ng(\theta_{-n}) = \frac{n(n-1)}{2} \theta'_n \quad \text{or} \quad g(\theta_{-n}) = \frac{n-1}{2} \theta'_n.$$

It is easily verified that the above expression for $g(\theta_{-n})$ is identical to the one given by Eq. (6.9).

Induction Step: Suppose that the hypothesis is true for all $k \leq K - 1$. Let $\theta = (\theta_1, \cdots, \theta_{K-1}, \theta_n, \cdots, \theta_n) \in \Theta^K$. Observe that any efficient queue σ for this profile must be such that $\sigma_i = i, i = 1, \ldots, K - 1$. *Budget balance* implies that

$$\sum_{i \in N} g(\theta_{-i}) = \sum_{i=1}^{K-1} (i-1)\theta_i + \frac{(n-K+1)(n+K-2)}{2} \theta'_n. \tag{6.10}$$

Let $i \in \{1, \ldots, K-1\}$. Define θ^i by

$$\theta^i_j = \begin{cases} \theta_j & \text{if } j < i, \\ \theta_{j+1} & \text{if } n > j \geq i, \\ \theta'_n & \text{otherwise.} \end{cases}$$

Thus, the first $i - 1$ elements of θ^i are the same as θ; for the next $n - i$ elements, the jth element of θ^i is the $(j + 1)$th element of θ; and the last element of θ^i is θ'_n. Observe that $\theta^i_j = \theta'_n$ for all $j \geq K - 1$ and hence $\theta^i \in \Theta^{K-1}$.
For $i = 1, \ldots, K - 1$, θ^i_{-n} is a permutation of θ_{-i}. Hence the symmetry of g implies that $g(\theta_{-i}) = g(\theta^i_{-n})$. Symmetry also implies that for any $i, j \in$

$\{K, \ldots, n\}$, $g(\theta_{-i}) = g(\theta_{-j})$. Hence, for $i = K, \ldots, n$, we can write $g(\theta_{-i}) = g(\theta_{-n})$. Therefore, we can write Eq. (6.10) as

$$\sum_{i=1}^{K-1} g(\theta^i_{-n}) + (n-K+1)g(\theta_{-n}) = \sum_{i=1}^{K-1}(i-1)\theta_i + \frac{(n-K+1)(n+K-2)}{2}\theta'_n.$$

Using the induction hypothesis on the profiles θ^i, we have

$$g(\theta^i_{-n}) = \sum_{r=1}^{K-2}\left(\frac{r-1}{n-2}\right)\theta^{i-1}_r + \frac{(n+K-4)(n-K+1)}{2(n-2)}\theta'_n.$$

Hence,

$$(n-K+1)g(\theta_{-n}) = \sum_{i=1}^{K-1}(i-1)\theta_i - \sum_{i=1}^{K-1}\sum_{r=1}^{K-2}\left(\frac{r-1}{n-2}\right)\theta^i_r$$
$$+\left[\frac{(n-K+1)(n+K-2)}{2} - \frac{(K-1)(n+K-4)(n-K+1)}{2(n-2)}\right]\theta'_n.$$

This simplifies to

$$(n-K+1)g(\theta_{-n}) = \sum_{i=1}^{K-1}(i-1)\theta_i - \sum_{i=1}^{K-1}\sum_{r=1}^{K-2}\left(\frac{r-1}{n-2}\right)\theta^i_r$$
$$+\frac{(n-K+1)(n-K)(n+K-3)}{2(n-2)}\theta'_n.$$

Recall that $\theta^i_j = \theta_j, j = 1, \ldots, i-1$ and $\theta^i_j = \theta_{j+1}, j = i, \ldots, K-2$. Note, however, that the order in the efficient queue is the same for $i = 1, \ldots, K-2$ in the profiles θ and θ^i. Using these observations,

$$\sum_{r=1}^{K-2}\left(\frac{r-1}{n-2}\right)\theta^i_r = \sum_{r=1}^{i-1}\left(\frac{r-1}{n-2}\right)\theta_r + \sum_{r=i}^{K-2}\left(\frac{r-1}{n-2}\right)\theta_{r+1}.$$

Note that θ_r appears in exactly $K-2$ of the profiles of the type θ^i. In particular, θ_r appears in all profiles except for θ^r. For $i = 1, \ldots, r-1$, θ_r is the waiting cost of agent $r-1$ (in the profile θ^i); in the other profiles, it is the waiting cost of agent r. It thus follows that the coefficient of θ_r in the expression

$$\sum_{i=1}^{K-1}(i-1)\theta_i - \sum_{i=1}^{K-1}\sum_{j=1}^{K-2}\left(\frac{j-1}{n-2}\right)\theta^i_j$$

is

$$r-1-\frac{(r-2)(r-1)}{n-2}-\frac{(K-r-1)(r-1)}{n-2}=\frac{(r-1)(n-K+1)}{n-2}.$$

We thus have

$$(n-K+1)g(\theta_{-n})=\sum_{r=1}^{K-1}\frac{(r-1)(n-K+1)}{n-2}\theta_r+\frac{(n-K+1)(n-K)(n+K-3)}{2(n-2)}\theta'_n.$$

Dividing across by $(n-K+1)$ establishes the induction step.

We can now establish Eq. (6.8) by considering the case $k=n$. In this case, the second term drops out of Eq. (6.9), and the expression is exactly what we want to establish. Since $g(\theta'_{-n})$ does not depend on θ'_n, it follows that Eq. (6.8) also applies to any profile where the waiting costs of agents in $N \setminus \{n\}$ have the same ordinal ranking. Observe that Eq. (6.7) reduces to Eq. (6.8).

Now let $i \neq n$. Rename the agents so that agents i and n interchange names with the others retaining their original names. We can do the same argument, and at the end interchange names again, to conclude that $g(\theta'_{-i})$ is given by Eq. (6.7). To complete the proof, we need to consider the case when $\theta'_1 \geq \cdots \geq \theta'_n$. In this case, the same proof goes through except that we have to use a tie-breaking rule to select an efficient queue. It can be verified that no matter what tie-breaking rule is used, $g(\theta'_{-i})$ will still be given by Eq. (6.7). □

Remark 6.9 Holmström's result can only be applied to *single-valued rules,* which requires that at each problem, a rule should choose a unique allocation from the feasible set of allocations. On the other hand, Kayi and Ramaekers (2010, in press) work with *multi-valued rules* (or *correspondences*), which allows the possibility of choosing a subset of allocations at each problem. Thus, they cannot use Holmström's result and the proof becomes complicated.

One might ask naturally how much generality is lost in our approach. In this regard, we note here that the multi-valued rule which selects the set of efficient queues at each profile is "essentially" single-valued. The efficient queue is unique at all profiles where no two agents have identical waiting costs. We can show that the set of all such profiles is an open and dense set in $\mathbf{R}^n_+$, and thus the efficient queue is generically unique. Hence, imposing a tie-breaking rule on profiles where the efficient queue is not unique does not amount to a significant loss of generality. Moreover, we emphasize that our result applies for any choice of tie-breaking rule and agents with the same waiting cost end up with the same utility, no matter which tie-breaking rule is used to select the efficient queue.

6.3.2 *Pivotal Rules*

By generalizing the idea of the pivotal rule, Mitra and Mutuswami (2011) introduce the k-pivotal rules in the context of queueing problems.

k-Pivotal rules, μ^k: For all $k \in \{1, \ldots, n\}$, all $N \in \mathcal{N}$, and all $\theta \in \mathcal{Q}^N$, $\mu^k(\theta) = (\sigma^k(\theta), t^k(\theta))$ is defined as $\sigma^k(\theta) \in \mathit{Eff}(\theta)$, and for all $i \in N$,

$$t_i^k(\theta) = \begin{cases} -\sum_{j:\sigma_i^k(\theta)<\sigma_j^k(\theta)\leq k} \theta_j & \text{if } \sigma_i^k(\theta) < k, \\ 0 & \text{if } \sigma_i^k(\theta) = k, \\ \sum_{j:k\leq\sigma_j^k(\theta)<\sigma_i^k(\theta)} \theta_j & \text{if } \sigma_i^k(\theta) > k. \end{cases}$$

The significance of the k-pivotal rules lies in the fact that they are immune to a particular form of group deviation. In fact, the k-pivotal rules are *weak group strategyproof* which requires that no coalition deviate in a manner benefiting all deviating members strictly.[6] *Pairwise strategyproofness* is *weak group strategyproofness* restricted to coalitions of size at most two. Moreover, Mitra and Mutuswami (2011) characterize the k-pivotal rules by imposing the axioms of *queue-efficiency, equal treatment of equals, pairwise strategyproofness,* and *weak linearity,* which is a technical condition requiring that transfers vary in a linear fashion whenever an agent changes her announcement in a manner which does not change the efficient queue. In addition, it can be shown that the k-pivotal rules also satisfy the normative criterion of *no-envy*.

First, we provide axiomatic characterization of two pivotal rules (see Sect. 2.4 for their definitions) from the class of anonymous VCG rules by additionally imposing the independence axioms introduced in Sect. 4.2. Once again, since we assume in this chapter that a rule is single-valued, we need to modify their definitions accordingly.

Pivotal rule, μ^P: For all $N \in \mathcal{N}$ and all $\theta \in \mathcal{Q}^N$, $\mu^P(\theta) = (\sigma^P(\theta), t^P(\theta))$ is defined as $\sigma^P(\theta) \in \mathit{Eff}(\theta)$, and for all $i \in N$,

$$t_i^P(\theta) = -\sum_{j\in F_i(\sigma^P(\theta))} \theta_j. \tag{6.11}$$

Reward-based pivotal rule, μ^R: For all $N \in \mathcal{N}$ and all $\theta \in \mathcal{Q}^N$, $\mu^R(\theta) = (\sigma^R(\theta), t^R(\theta))$ is defined as $\sigma^R(\theta) \in \mathit{Eff}(\theta)$, and for all $i \in N$,

$$t_i^R(\theta) = \sum_{j\in P_i(\sigma^R(\theta))} \theta_j. \tag{6.12}$$

---

[6]Formally, *weak group strategyproofness* requires that for all $S \subset N$ and all $\theta, \theta' \in \mathcal{Q}^N$ such that $\theta_i = \theta'_i$ for all $i \in N \backslash S$, it is not the case that $u_i(\mu_i(\theta'); \theta_i) > u_i(\mu_i(\theta); \theta_i)$ for all $i \in S$. We have *pairwise strategyproofness* if we also require $|S| \leq 2$.

Remark 6.10 The pivotal rule is the n-pivotal rule while the reward-based pivotal rule is the 1-pivotal rule.

We use two independence axioms introduced in Sect. 4.2 to characterize the pivotal and the reward-based pivotal mechanisms, which are based on the idea that if an agent's waiting cost changes, then some agents should remain unaffected. *Independence of preceding costs* (Maniquet 2003) requires that an increase in an agent's waiting cost should not affect her followers. This reflects the idea that when an agent's waiting cost increases, while the aggregate waiting cost increases, this cannot be attributed to the agent's followers. On the other hand, *independence of following costs* (Chun 2006a) requires that a decrease in an agent's waiting cost should not affect the predecessors. This reflects a similar idea that when an agent's cost decreases, while the aggregate waiting cost decreases, this cannot be attributed to the agent's predecessors.

Remark 6.11 It is obvious that the pivotal rule satisfies *independence of preceding costs* and the reward-based pivotal rule satisfies *independence of following costs.* Our characterization results still hold even if *independence of preceding costs* and *independence of following costs* are weakened by imposing an additional assumption that the efficient queue remains the same before and after a change in an agent's waiting cost.

Our last axiom is a mild regularity condition on the transfers saying that if all agents have zero waiting costs, then their transfers must add up to zero.

Weak budget balance: For all $N \in \mathcal{N}$ and all $\theta \in \mathcal{Q}^N$, if $\theta = (0, \ldots, 0)$, then $\sum_{i \in N} t_i(\theta) = 0$.

The following theorem characterizes the pivotal and the reward-based pivotal rules.

Theorem 6.2

1. *The pivotal rule is the only rule satisfying queue-efficiency, equal treatment of equals, strategyproofness, independence of preceding costs, and weak budget balance.*
2. *The reward-based pivotal rule is the only rule satisfying queue-efficiency, equal treatment of equals, strategyproofness, independence of following costs, and weak budget balance.*

Proof It is obvious that the pivotal rule satisfies *queue-efficiency, equal treatment of equals, strategyproofness, independence of preceding costs,* and *weak budget balance,* whereas the reward-based pivotal rule satisfies *queue-efficiency, equal treatment of equals, strategyproofness, independence of following costs,* and *weak budget balance.* Hence, it remains to show the converse statement.

Let $N \in \mathcal{N}$ and $\theta \in \mathcal{Q}^N$. Suppose without loss of generality that $\theta_1 \geq \theta_2 \geq \cdots \geq \theta_n$ and that for each $i \in N$, $\sigma_i = i$. By Proposition 6.1, the transfer can be expressed as, for all $i \in N$,

$$t_i(\theta) = -\sum_{j \in F_i(\sigma(\theta))} \theta_j + g(\theta_{N\setminus\{i\}}). \tag{6.13}$$

Let μ be a rule satisfying *queue-efficiency, equal treatment of equals, strategyproofness, independence of preceding costs,* and *weak budget balance,* and for all $\theta \in \mathcal{Q}^N$, $\mu(\theta) = (\sigma(\theta), t(\theta))$. By Eq. (6.13), $t_n(\theta) = g(\theta_{N\setminus\{n\}})$. By *independence of preceding costs,* for all θ, $\theta' \in \mathcal{Q}^N$ such that $\theta'_i \geq \theta_i$ for all $i \neq n$, and $\theta'_n = \theta_n$, $t_n(\theta) = t_n(\theta')$. This implies that there exists $c \in \mathbf{R}$ such that for all $\theta_{N\setminus\{n\}}$, $g(\theta_{N\setminus\{n\}}) = c$. By *weak budget balance,* $\sum_{i \in N} t_i(\theta) = nc = 0$ at $\theta = (0, \ldots, 0)$ and hence $c = 0$. Altogether, we conclude that for all $i \in N$, $t_i(\theta) = -\sum_{j \in F_i(\sigma(\theta))} \theta_j = t_i^P(\theta)$, as desired.

Now let μ be a rule satisfying *queue-efficiency, equal treatment of equals, strategyproofness, independence of following costs,* and *weak budget balance,* and for all $\theta \in \mathcal{Q}^N$, $\mu(\theta) = (\sigma(\theta), t(\theta))$. Then, $t_1(\theta) = -\sum_{j \in F_1(\sigma(\theta))} \theta_j + g(\theta_{N\setminus\{1\}})$. By *independence of following costs,* for all θ, $\theta' \in \mathcal{Q}^N$ such that $\theta'_i \leq \theta_i$ for all $i \neq 1$, and $\theta'_1 = \theta_1$, $t_1(\theta) = t_1(\theta')$. This implies that there exists $c \in \mathbf{R}$ such that for all $\theta_{N\setminus\{1\}}$, $g(\theta_{N\setminus\{1\}}) = \sum_{j \in F_1(\sigma(\theta))} \theta_j + c$. By *weak budget balance,* $\sum_{i \in N} t_i(\theta) = nc = 0$ at $\theta = (0, \ldots, 0)$ and hence $c = 0$. Altogether, we again conclude that for all $i \in N$,

$$\begin{aligned} t_i(\theta) &= -\sum_{j \in F_i(\sigma(\theta))} \theta_j + g(\theta_{N\setminus\{i\}}) \\ &= -\sum_{j \in F_i(\sigma(\theta))} \theta_j + \sum_{j \in N\setminus\{i\}} \theta_j \\ &= \sum_{j \in P_i(\sigma(\theta))} \theta_j \\ &= t_i^R(\theta), \end{aligned}$$

as desired. □

In the following two remarks, we check the independence of our axioms used to characterize the pivotal and the reward-based pivotal rules.

Remark 6.12 The pivotal rule (σ, t^P):

1. Dropping *queue-efficiency*: Let $N = \{1, 2\}$. Suppose that for all $\theta \in \mathcal{Q}^N$ and all $i \in N$, $\sigma_i(\theta) = i$ and the transfers are given by

$$t_i(\theta) = \begin{cases} -\theta_2 \text{ if } i = 1, \\ 0 \text{ if } i = 2. \end{cases}$$

This rule satisfies *equal treatment of equals, strategyproofness, independence of preceding costs,* and *weak budget balance.*

2. Dropping *equal treatment of equals*: Consider the rule $(\sigma, \hat{t}^P)$ where for all $\theta \in \mathcal{Q}^N$ and all $i \in N$, $\sigma(\theta) \in \mathit{Eff}(\theta)$, $\hat{t}_i^P(\theta) = t_i^P(\theta) + a_i$, $\sum_{i \in N} a_i = 0$, and there exists at least a pair $i, j \in N$ such that $a_i \neq a_j$. This rule satisfies *queue-efficiency, strategyproofness, independence of preceding costs,* and *weak budget balance.*
3. Dropping *strategyproofness*: The minimal transfer rule (Maniquet 2003) satisfies *queue-efficiency, equal treatment of equals, independence of preceding costs,* and *weak budget balance.*
4. Dropping *independence of preceding costs*: The reward-based pivotal rule satisfies *queue-efficiency, equal treatment of equals, strategyproofness,* and *weak budget balance.*
5. Dropping *weak budget balance*: Consider the rule $(\sigma, \bar{t}^P)$ where for all $\theta \in \mathcal{Q}^N$ and all $i \in N$, $\sigma(\theta) \in \mathit{Eff}(\theta)$, $\bar{t}_i^P(\theta) = t_i^P(\theta) + c$, and $c \neq 0$. This rule satisfies *queue-efficiency, equal treatment of equals, strategyproofness,* and *independence of preceding costs.*

Remark 6.13 The reward-based pivotal rule (σ, t^R):

1. Dropping *queue-efficiency*: Let $N = \{1, 2\}$. Suppose that for all $\theta \in \mathcal{Q}^N$ and all $i \in N$, $\sigma_i(\theta) = i$ and the transfers are given by

$$t_i(\theta) = \begin{cases} 0 & \text{if } i = 1, \\ \theta_1 & \text{if } i = 2. \end{cases}$$

This rule satisfies *equal treatment of equals, strategyproofness, independence of following costs,* and *weak budget balance.*

2. Dropping *equal treatment of equals*: Consider the rule $(\sigma, \hat{t}^R)$ where for all $\theta \in \mathcal{Q}^N$ and all $i \in N$, $\sigma(\theta) \in \mathit{Eff}(\theta)$, $\hat{t}_i^R(\theta) = t_i^R(\theta) + b_i$, $\sum_{i \in N} b_i = 0$, and there exists at least a pair $i, j \in N$ such that $b_i \neq b_j$. This rule satisfies *queue-efficiency, strategyproofness, independence of following costs,* and *weak budget balance.*
3. Dropping *strategyproofness*: The maximal transfer rule (Chun 2006a) satisfies *queue-efficiency, equal treatment of equals, independence of following costs,* and *weak budget balance.*
4. Dropping *independence of following costs*: The pivotal rule satisfies *queue-efficiency, equal treatment of equals, strategyproofness,* and *weak budget balance.*
5. Dropping *weak budget balance*: Consider the rule $(\sigma, \bar{t}^R)$ where for all $\theta \in \mathcal{Q}^N$ and all $i \in N$, $\sigma(\theta) \in \mathit{Eff}(\theta)$, $\bar{t}_i^R(\theta) = t_i^R(\theta) + c$, and $c \neq 0$. This rule satisfies *queue-efficiency, equal treatment of equals, strategyproofness,* and *independence of following costs.*

Remark 6.14 Equal treatment of equals and *weak budget balance* together imply the following regularity condition on the transfers. A rule μ satisfies the *zero transfer condition* if $u_i(\mu_i(\theta); \theta_i) = 0$ for all $i \in N$ whenever $\theta = (0, \ldots, 0)$. Both the pivotal and the reward-based pivotal rules satisfy the *zero transfer condition.* In the characterization of the pivotal rule, we can replace *weak budget balance* by the *zero transfer condition*, that is, the pivotal rule is the only rule that satisfies *queue-efficiency, equal treatment of equals, strategyproofness, independence of preceding costs,* and the *zero transfer condition*. However, if we use the *zero transfer condition* to characterize the reward-based pivotal rule, then we can eliminate both *weak budget balance* and *equal treatment of equals*. Hence we can show that the reward-based pivotal rule is the only rule that satisfies *queue-efficiency, strategyproofness, independence of following costs,* and the *zero transfer condition*.

Now we generalize independence axioms to characterize all k-pivotal rules. Given $k = 1, \ldots, n$, *k-independence of preceding costs* requires that if a waiting cost of an agent whose position is no later than k increases, then her followers whose position is also no later than k should not be affected. On the other hand, *k-independence of following costs* requires that if a waiting cost of an agent whose position is no earlier than k decreases, then her predecessors whose position is also no earlier than k should not be affected. In other words, given a profile, *k-independence of preceding costs* applies to all agents whose queue position is no later than k and *k-independence of following costs* applies to all agents whose queue position is no earlier than k. We define the axioms formally now.

k-independence of preceding costs: For all $N \in \mathcal{N}$, all $\theta, \theta' \in \mathcal{Q}^N$, and all $i \in N$, if for all $j \in N \setminus \{i\}$, $\theta'_j = \theta_j$, $\theta'_i > \theta_i$, and $\sigma_i(\theta) \leq k$, then for all $\ell \in F_i(\sigma(\theta))$ such that $\sigma_\ell(\theta) \leq k$, $u_\ell(\mu_\ell(\theta'); \theta'_\ell) = u_\ell(\mu_\ell(\theta); \theta_\ell)$.

k-independence of following costs: For all $N \in \mathcal{N}$, all $\theta, \theta' \in \mathcal{Q}^N$, and all $i \in N$, if for all $j \in N \setminus \{i\}$, $\theta'_j = \theta_j$, $\theta'_i < \theta_i$, and $\sigma_i(\theta) \geq k$, then for all $\ell \in P_i(\sigma(\theta))$ such that $\sigma_\ell(\theta) \geq k$, $u_\ell(\mu_\ell(\theta'); \theta'_\ell) = u_\ell(\mu_\ell(\theta); \theta_\ell)$.

Observe that *n-independence of preceding costs* is identical to *independence of preceding costs*, while 1*-independence of following costs* is the same as *independence of following costs*. Also, note that both *n-independence of following costs* and 1*-independence of preceding costs* are redundant. The characterization result for k-pivotal rule is as follows.

Theorem 6.3 *A rule satisfies queue-efficiency, equal treatment of equals, strategyproofness, k-independence of preceding costs, k-independence of following costs, and weak budget balance if and only if it is a k-pivotal rule.*

Proof It is easy to verify that the k-pivotal rules satisfy *queue-efficiency, equal treatment of equals, strategyproofness, k-independence of preceding costs, k-independence of following costs,* and *weak budget balance*. To prove the converse statement, let $\mu = (\sigma, t)$ be a rule satisfying the six axioms. By Proposition 6.1, we know that μ is anonymous, and hence, we can write the transfers as

$$t_i(\theta) = - \sum_{j \in F_i(\sigma(\theta))} \theta_j + f(\theta_{N \setminus \{i\}}). \tag{6.14}$$

Without loss of generality, suppose that $\theta_1 \geq \cdots \geq \theta_n$. By *equal treatment of equals,* we may assume that $\sigma_k(\theta) = k$. By *k-independence of preceding costs,* it follows that $t_k(\theta)$ cannot depend on $(\theta_1, \ldots, \theta_{k-1})$ and we have

$$f(\theta_{N\setminus\{k\}}) = f(\theta_{k+1}, \ldots, \theta_n). \tag{6.15}$$

On the other hand, by *k-independence of following costs,* it follows that

$$f(\theta_{N\setminus\{k\}}) = \sum_{j=k+1}^{n} \theta_j + g(\theta_1, \ldots, \theta_{k-1}). \tag{6.16}$$

Hence, it follows that $g(\theta_1, \ldots, \theta_{k-1})$ is a constant and

$$f(\theta_{N\setminus\{k\}}) = \sum_{j=k+1}^{n} \theta_j + c_k \tag{6.17}$$

where c_k is a player-specific constant.

Since the f function is the same for all agents, it follows that for all $\theta \in \mathcal{Q}^N$ and all $i \in N$,

$$f(\theta_{N\setminus\{i\}}) = \sum_{j:\sigma_j(\theta_{N\setminus\{i\}})\geq k} \theta_j + c_i$$

By *weak budget balance* and *equal treatment of equals,* for all $i \in N$, $c_i = 0$ and the result follows by substituting for $f(\theta_{N\setminus\{i\}})$. □

Remark 6.15 We now show that all axioms are necessary for the characterization.

1. Dropping *queue-efficiency*: Let $N = \{1, 2\}$. For $k = 1$, the example in Remark 6.12 (1) and for $k = 2$, the example in Remark 6.13(1) satisfy *equal treatment of equals, strategyproofness, k-independence of preceding costs, k-independence of following costs,* and *weak budget balance.*
2. Dropping *equal treatment of equals*: Let $N = \{1, 2\}$. For $k = 1$, the example in Remark 6.12 (2) and for $k = 2$, the example in Remark 6.13(2) satisfy *queue-efficiency, strategyproofness, k-independence of preceding costs, k-independence of following costs,* and *weak budget balance.*
3. Dropping *strategyproofness*: Let $k = 1, \ldots, n$ be given. Assume that $\theta_1 \geq \theta_2 \geq \cdots \geq \theta_n$ and also assume that $\sigma_i = i$ for all $i \in N$. The transfers are:

$$t_i(\theta) = \begin{cases} -\sum_{j=i+1}^{k} \frac{1}{2}\theta_j - \frac{n-k}{2}\theta_k \text{ if } i = 1, \\ \frac{i-1}{2}\theta_i - \sum_{j=i+1}^{k} \frac{1}{2}\theta_j - \frac{n-k}{2}\theta_k \text{ if } 1 < i \leq k, \\ \frac{k-1}{2}\theta_k + \sum_{j=k+1}^{i-1} \frac{1}{2}\theta_j - \frac{n-i}{2}\theta_i \text{ if } k+1 < i < n, \\ \frac{k-1}{2}\theta_k + \sum_{j=k+1}^{i-1} \frac{1}{2}\theta_j \text{ if } i = n. \end{cases}$$

This rule satisfies *queue-efficiency, equal treatment of equals, k-independence of preceding costs, k-independence of following costs,* and *weak budget balance.*

4. Dropping *k-independence of preceding costs*: The rule (σ, t) where $\sigma(\theta) \in Eff(\theta)$ and for all $i \in N$, $t_i(\theta) = t_i^k(\theta) + \frac{1}{2}\max_{j \neq i} \theta_j$ satisfies *queue-efficiency, equal treatment of equals, strategyproofness, k-independence of following costs,* and *weak budget balance.*
5. Dropping *k-independence of following costs*: The rule (σ, t) where $\sigma(\theta) \in Eff(\theta)$ and for all $i \in N$, $t_i(\theta) = t_i^k(\theta) - \frac{1}{2}\min_{j \neq i} \theta_j$ satisfies *queue-efficiency, equal treatment of equals, strategyproofness, k-independence of preceding costs,* and *weak budget balance.*
6. Dropping *weak budget balance*: The rule (σ, t) where $\sigma(\theta) \in Eff(\theta)$ and for all $i \in N$, $t_i(\theta) = t_i^k(\theta) + c$ where $c \neq 0$ satisfies *queue-efficiency, equal treatment of equals, strategyproofness, k-independence of preceding costs,* and *k-independence of following costs.*

Remark 6.16 When $k = n$, note that we can dispense with *n-independence of following costs* because Eq. (6.15) directly implies that the f function must be a constant. Similarly, when $k = 1$, we can dispense with 1-*independence of preceding costs*, because we get the result directly from Eq. (6.16). However, we need both *k-independence of preceding costs* and *k-independence of following costs* for other values of k.

Remark 6.17 We can also characterize the k-pivotal rules by imposing a "normalization" axiom. Specifically, a rule satisfies *k-normalization* if for all $\theta \in \mathcal{Q}^N$ and all $i \in N$, $\sigma_i(\theta) = k$ implies $t_i(\theta) = 0$. It is not difficult to show that *queue-efficiency, equal treatment of equals, strategyproofness,* and *k-normalization* characterize the k-pivotal rules.

6.4 Concluding Remarks

A natural question that arises is whether the characterizations can be extended to scheduling problems where agents have different processing times (which are known) and different waiting costs (which are private information). This is not easy because our characterizations rely on Proposition 6.1 which characterize the set of *queue-efficient* and *strategyproof* rules satisfying *equal treatment of equals.* When agents have different processing times as well as different waiting costs, it is not clear what *equal treatment of equals* means and whether an equivalent to Proposition 6.1 can be obtained. This is an undoubtedly interesting open question.

References

Barberà, S. (2011). Strategyproof social choice. In K. J. Arrow, A. K. Sen, & K. Suzumura (Eds.), *Handbook of social choice and welfare* (Vol. 2, pp. 731–831). Amsterdam: Elsevier BV.

Chun, Y. (2006a). A pessimistic approach to the queueing problem. *Mathematical Social Sciences, 51*, 171–181.

Chun, Y. (2006b). No-envy in queueing problems. *Economic Theory, 29*, 151–162.

Chun, Y., Mitra, M., & Mutuswami, S. (2011). No-envy, egalitarian equivalence and strategyproofness in queueing problems. Mimeo, Seoul National University.

Chun, Y., Mitra, M., & Mutuswami, S. (2014). Characterizations of pivotal mechanisms in the queueing problem. *Mathematical Social Sciences, 72*, 62–66.

Chun, Y., Mitra, M., & Mutuswami, S. (in press). A characterization of the symmetrically balanced VCG rule in the queueing problem. Games and Economic Behavior.

Clarke, E. H. (1971). Multi-part pricing of public goods. *Public Choice, 11*, 17–33.

Foley, D. (1967). Resource allocation and the public sector. *Yale Economic Essays, 7*, 45–98.

Groves, T. (1973). Incentives in teams. *Econometrica, 41*, 617–631.

Hashimoto, K., & Saitoh, H. (2012). Strategy-proof and anonymous rules in queueing problems: A relationship between efficiency and equity. *Social Choice and Welfare, 38*, 473–480.

Holmström, B. (1979). Groves' schemes on restricted domains. *Econometrica, 47*, 1137–1144.

Kayi, C., & Ramaekers, E. (2010). Characterizations of Pareto-efficient, fair, and strategy-proof allocation rules in queueing problems. *Games and Economic Behavior, 68*, 220–232.

Kayi, C., & Ramaekers. E. (in press). Corrigendum to Characterizations of Pareto-efficient, fair, and strategy-proof allocation rules in queueing problems [Games and Economic Behavior 68:220–232]. Games and Economic Behavior.

Maniquet, F. (2003). A characterization of the Shapley value in queueing problems. *Journal of Economic Theory, 109*, 90–103.

Mitra, M. (2001). Mechanism design in queueing problems. *Economic Theory, 17*, 277–305.

Mitra, M., & Mutuswami, S. (2011). Group strategyproofness in queueing models. *Games and Economic Behavior, 72*, 242–254.

Pápai, S. (2003). Groves sealed bid auctions of heterogeneous objects with fair prices. *Social Choice and Welfare, 20*, 371–385.

Suijs, J. (1996). On incentive compatibility and budget balancedness in public decision making. *Economic Design, 2*, 193–209.

Thomson, W. (2013). Strategy-proof allocation rules. Book manuscript, University of Rochester.

Vickrey, W. (1961). Counterspeculation, auctions and competitive sealed tenders. *Journal of Finance, 16*, 8–37.

Chapter 7
Strategyproofness and Egalitarian Equivalence

7.1 Introduction

Egalitarian equivalence (Pazner and Schmeidler 1978) requires that for each preference profile, there should be a reference bundle such that each agent is indifferent between her bundle and the reference bundle. Like no-envy, an attractive feature of *egalitarian equivalence* is that it is an ordinal concept and makes no interpersonal utility comparisons. Since the reference bundle is common, it is easy to see that it satisfies *equal treatment of equals* or horizontal equity. It has also been studied in many contexts (Demange 1984; Dutta and Vohra 1993; Thomson 1990; Yengin 2011, 2012). In this chapter, we investigate the implications of *egalitarian equivalence* in the context of queueing problems.

As in Chap. 6, our starting point is the classic result of Holmström (1979) which implies that in our context, a rule satisfies *queue-efficiency* and *strategyproofness* if and only if it is a VCG rule.[1] *Queue-efficiency* requires that the selected queue should minimize the aggregate waiting cost, and *strategyproofness* requires that an agent should not have an incentive to misrepresent her waiting cost no matter what she believes other agents to be doing. It follows that imposing an additional axiom of *egalitarian equivalence* gives us a subset of VCG rules.

First, we present a complete characterization of the family of rules satisfying *queue-efficiency, strategyproofness,* and *egalitarian equivalence*.[2] To understand our characterization, note that a reference bundle for a (preference) profile is a queue position along with a corresponding transfer. The reference bundle can change across profiles. We show that if we impose *queue-efficiency* and *strategyproofness* along with *egalitarian equivalence,* then there is effectively only one degree of

[1]The family of VCG rules are due to Vickrey (1961), Clarke (1971), and Groves (1973).

[2]The same issue was addressed for the problem of allocating heterogeneous objects by Yengin (2011, 2012).

© Springer International Publishing Switzerland 2016

Y. Chun, *Fair Queueing*, Studies in Choice and Welfare,
DOI 10.1007/978-3-319-33771-5_7

freedom in choosing the reference bundle. In particular, once a queue position is selected for a profile, then we must select the same position for all profiles. Furthermore, the transfers are determined for all profiles up to a type-independent constant.

We go on to show that none of these rules satisfies *budget balance*. However, we obtain a possibility result with the weaker notion of *feasibility* which allows a rule to run budget surpluses but not deficits. We characterize all rules satisfying *queue-efficiency, strategyproofness, egalitarian equivalence,* and *feasibility*. This class restricts the reference bundle to choose only the first queue position. However, there cannot be an upper bound that can be placed on the resulting budget surplus.

Another desirable property of a rule is immunity to manipulations by groups of agents. *Weak group strategyproofness* requires that it is not possible for a group of agents to manipulate their reports in a manner which makes all of them strictly better off. We show that if there are three or more agents, then we cannot find rules satisfying *queue-efficiency, weak group strategyproofness,* and *egalitarian equivalence*.

The contrast between the results obtained here and the results obtained using *no-envy* is striking. With regard to *budget balance,* it is shown in Sect. 6.3.1 that the symmetrically balanced VCG rule satisfies *no-envy, strategyproofness,* and *budget balance.*[3] With regard to *weak group strategyproofness,* as shown in Sect. 6.3.2, the k-pivotal rules satisfy *no-envy* and *weak group strategyproofness*. The rather sharp contrast leads us to the incompatibility between no-envy and egalitarian equivalence in Sect. 5.5.

The chapter is organized as follows.[4] Section 7.2 presents the characterization result on *queue-efficient, strategyproof,* and *egalitarian equivalent* rules. Section 7.3 discusses the consequence of additionally imposing *budget balance* or the weaker requirement of *feasibility*. Section 7.4 shows that *strategyproofness* cannot be strengthened to *weak group strategyproofness* and discusses the relationship between *no envy* and *egalitarian equivalence* together with *strategyproofness*. Concluding remarks follow in Sect. 7.5.

7.2 Strategyproofness and Egalitarian Equivalence

As in Chap. 6, we begin with the classic result of Holmström (1979) which implies that in the context of queueing problems, a rule satisfies *queue-efficiency* and *strategyproofness* if and only if it is a VCG rule. *Queue-efficiency* requires that the selected queue should minimize the aggregate waiting cost so that agents receive their service in the non-increasing order of their waiting costs. *Strategyproofness,* as studied in Chap. 6, requires that an agent cannot strictly gain by misrepresenting her

---

[3]As shown in Proposition 5.1, *no-envy* implies *queue-efficiency*.

[4]This chapter is based mainly on Chun et al. (2014).

waiting cost, no matter what she believes other agents to be doing. By imposing an additional axiom, we have a subset of VCG rules.

Also, as in Chap. 6, to use Holmström's result, we assume that a rule is single-valued and therefore, denoted by μ. It implicitly assumes the existence of a tie-breaking rule as explained in Remark 6.1. For all $N \in \mathscr{N}$ and all $\theta \in \mathscr{Q}^N$, we abuse the notation and write $\mu(\theta) = (\sigma(\theta), t(\theta))$. First, we recall the definition of the VCG rules given in Eq. (6.2).

VCG rule associated with g_i, μ^{g_i}: For all $N \in \mathscr{N}$ and all $\theta \in \mathscr{Q}^N$, $\mu^{g_i}(\theta) = (\sigma(\theta), t(\theta))$ is defined as: $\sigma(\theta) \in \mathit{Eff}(\theta)$ and for all $i \in N$,

$$t_i(\theta) = -\sum_{j \in F_i(\sigma)} \theta_j + g_i(\theta_{N\setminus\{i\}}).$$

Egalitarian equivalence (Pazner and Schmeidler 1978) is based on the idea that at some allocation, if each agent enjoys the same utility between her bundle and the common reference bundle, then the allocation can be evaluated as reasonable. A rule satisfies *egalitarian equivalence* if for each profile there exists a reference bundle such that each agent is indifferent between her bundle (for the profile) and the reference bundle (for the profile).

Egalitarian equivalence: For all $N \in \mathscr{N}$ and all $\theta \in \mathscr{Q}^N$, there exists a reference bundle (σ_0, t_0) such that for all $i \in N$, $u_i(\mu_i(\theta); \theta_i) = u_i(\sigma_0, t_0; \theta_i)$.

We use the following lemma to prove our main theorem. It shows that for a VCG rule to be *egalitarian equivalent*, $g_i(\theta_{N\setminus\{i\}})$ must be affine with the coefficients of θ_j, $j \neq i$, being pinned down by the choice of σ_0.

Lemma 7.1 *A rule μ satisfies queue-efficiency, strategyproofness, and egalitarian equivalence only if it is a VCG rule and there exists $\sigma_0 \in \{1, \ldots, n\}$ and $c \in \mathbf{R}$ such that for all $i \in N$ and all $\theta_{N\setminus\{i\}}$,*

$$g_i(\theta_{N\setminus\{i\}}) = \sum_{j \neq i} (\sigma_0 - \sigma_j(\theta_{N\setminus\{i\}}))\theta_j + c \tag{7.1}$$

where $\mu(\theta) = (\sigma(\theta), t(\theta))$.

Proof Let $N \in \mathscr{N}$, $\theta \in \mathscr{Q}^N$, and $(\sigma_0(\theta), t_0(\theta))$ the corresponding reference bundle. Suppose that the efficient queue (for the profile θ) is such that for all $i \in N$, $\sigma_i(\theta) = i$. By *queue-efficiency*, this implies that $\theta_1 \geq \theta_2 \geq \cdots \geq \theta_n$. To satisfy *queue-efficiency, strategyproofness,* and *egalitarian equivalence* together, the following condition must hold: for all $i \in N$,

$$-(\sigma_i(\theta) - 1)\theta_i - \sum_{j \in F_i(\sigma)} \theta_j + g_i(\theta_{N\setminus\{i\}}) = -(\sigma_0(\theta) - 1)\theta_i + t_0(\theta).$$

The left-hand side of the above expression is the utility from a VCG mechanism, and the right-hand side is the utility from the *egalitarian equivalence* requirement. We can rewrite as

$$t_0(\theta) = (\sigma_0(\theta) - \sigma_i(\theta))\theta_i - \sum_{j \in F_i(\sigma)} \theta_j + g_i(\theta_{N\setminus\{i\}}). \tag{7.2}$$

Choose two agents i and $i+1$. Noting that $\sigma_i(\theta) = i$ for all i, we have, using Eq. (7.2),

$$(\sigma_0(\theta) - i)\theta_i - \sum_{j>i} \theta_j + g_i(\theta_{N\setminus\{i\}})$$
$$= (\sigma_0(\theta) - i - 1)\theta_{i+1} - \sum_{j>i+1} \theta_j + g_{i+1}(\theta_{N\setminus\{i+1\}}).$$

Hence,

$$(\sigma_0(\theta) - i)\theta_i + g_i(\theta_{N\setminus\{i\}}) = (\sigma_0(\theta) - i)\theta_{i+1} + g_{i+1}(\theta_{N\setminus\{i+1\}}). \tag{7.3}$$

Since g_i does not depend on θ_i and g_{i+1} does not depend on θ_{i+1}, we can write the two functions as follows:

$$g_i(\theta_{N\setminus\{i\}}) = (\sigma_0(\theta) - i)\theta_{i+1} + f_{i,i+1}(\theta_{N\setminus\{i,i+1\}}), \tag{7.4}$$
$$g_{i+1}(\theta_{N\setminus\{i+1\}}) = (\sigma_0(\theta) - i)\theta_i + f_{i,i+1}(\theta_{N\setminus\{i,i+1\}}). \tag{7.5}$$

Note that Eq. (7.3) implies that the same function $f_{i,i+1}(\theta_{N\setminus\{i,i+1\}})$ appears in Eqs. (7.4) and (7.5).[5] Before proceeding further, we note one important observation: Eqs. (7.4) and (7.5) together imply that $\sigma_0(\theta)$ must be independent of θ_i and θ_{i+1} since the left-hand side of Eqs. (7.4) and (7.5) is independent of θ_i and θ_{i+1}, respectively. Since this applies for all $i = 1, \ldots, n-1$, it follows that $\sigma_0(\theta)$ is independent of θ, in effect, a constant. We thus drop the dependence on θ from now on.

Choosing the agents $i+1$ and $i+2$ (note that these are also "adjacent" agents in the efficient queue) and doing the same analysis, we obtain the analogue of Eq. (7.4):

$$g_{i+1}(\theta_{N\setminus\{i+1\}}) = (\sigma_0 - i - 1)\theta_{i+2} + f_{i+1,i+2}(\theta_{N\setminus\{i+1,i+2\}}). \tag{7.6}$$

[5]If $n = 2$, then $f_{i,i+1}(\theta_{N\setminus\{i,i+1\}}) = c$. Therefore, we have the desired expression directly from Eqs. (7.4) and (7.5).

Equating Eqs. (7.5) and (7.6) gives us a recursive relationship between $f_{i,i+1}$ and $f_{i+1,i+2}$:

$$f_{i+1,i+2}(\theta_{N\backslash\{i+1,i+2\}}) = (\sigma_0 - i)\theta_i - (\sigma_0 - i - 1)\theta_{i+2} + f_{i,i+1}(\theta_{N\backslash\{i,i+1\}}). \tag{7.7}$$

There are $n - 1$ functions of the type $f_{i,i+1}$ (for $i = 1, \ldots, n-1$) and $n - 2$ recursion relations. We will use the recursion relations to obtain an expression for $f_{n-1,n}$. We start by observing that $f_{i,i+1}(\theta_{N\backslash\{i,i+1\}})$ cannot depend on θ_i or θ_{i+1}. Similarly, $f_{i+1,i+2}(\theta_{N\backslash\{i+1,i+2\}})$ cannot depend on θ_{i+1} or θ_{i+2}. However, the term $-(\sigma_0 - i - 1)\theta_{i+2}$ appears on the right-hand side of Eq. (7.7). This implies that $f_{i,i+1}(\theta_{N\backslash\{i,i+1\}})$ must have $(\sigma_0(\theta) - i - 1)\theta_{i+2}$ as an additive term. In other words, we can write

$$f_{i,i+1}(\theta_{N\backslash\{i,i+1\}}) = (\sigma_0 - i - 1)\theta_{i+2} + h^i(\theta_{N\backslash\{i,i+1,i+2\}}).$$

Substituting in Eq. (7.7), we get

$$f_{i+1,i+2}(\theta_{N\backslash\{i+1,i+2\}}) = (\sigma_0 - i)\theta_i + h^i(\theta_{N\backslash\{i,i+1,i+2\}}). \tag{7.8}$$

Next, put $i + 1$ instead of i in Eq. (7.7) and substitute for $f_{i+1,i+2}(\theta_{N\backslash\{i+1,i+2\}})$ using Eq. (7.8). We have

$$\begin{aligned} f_{i+2,i+3}(\theta_{N\backslash\{i+2,i+3\}}) = {} & (\sigma_0 - i - 1)\theta_{i+1} - (\sigma_0 - i - 2)\theta_{i+3} \\ & + (\sigma_0 - i)\theta_i + h^i(\theta_{N\backslash\{i,i+1,i+2\}}). \end{aligned} \tag{7.9}$$

Observe that $f_{i+2,i+3}(\theta_{N\backslash\{i+2,i+3\}})$ cannot depend on θ_{i+3} but there is a term $-(\sigma_0 - i - 2)\theta_{i+3}$ on the right-hand side. It follows that

$$h^i(\theta_{N\backslash\{i,i+1,i+2\}}) = h^{i+1}(\theta_{N\backslash\{i,i+1,i+2,i+3\}}) + (\sigma_0 - i - 2)\theta_{i+3}. \tag{7.10}$$

Substituting Eq. (7.10) back in Eq. (7.9) gives us

$$\begin{aligned} f_{i+2,i+3}(\theta_{N\backslash\{i+2,i+3\}}) = {} & (\sigma_0 - i)\theta_i + (\sigma_0 - i - 1)\theta_{i+1} \\ & + h^{i+1}(\theta_{N\backslash\{i,i+1,i+2,i+3\}}). \end{aligned}$$

It is clear that this argument applies recursively. Starting with $i = 1$ and applying the above argument repeatedly, we can conclude that for $c \in \mathbf{R}$,

$$\begin{aligned} f_{n-1,n}(\theta_{N\backslash\{n-1,n\}}) &= \sum_{i=1}^{n-2}(\sigma_0 - i)\theta_i + h^{n-2}(\theta_{N\backslash N}) \\ &= \sum_{i=1}^{n-2}(\sigma_0 - i)\theta_i + c. \end{aligned} \tag{7.11}$$

Note that c is a type-independent arbitrary constant. We can now use the recursion relations equation (7.7) to conclude that for $i = 1, \ldots, n-1$[6]:

$$f_{i,i+1}(\theta_{N\setminus\{i,i+1\}}) = \sum_{j=1}^{i-1}(\sigma_0 - j)\theta_j + \sum_{j=i+1}^{n-1}(\sigma_0 - j)\theta_{j+1} + c. \tag{7.12}$$

Hence, using Eq. (7.4), for $i = 1, \ldots, n-1$,

$$\begin{aligned} g_i(\theta_{N\setminus\{i\}}) &= (\sigma_0 - i)\theta_{i+1} + \sum_{j=1}^{i-1}(\sigma_0 - j)\theta_j + \sum_{j=i+1}^{n-1}(\sigma_0 - j)\theta_{j+1} + c \\ &= \sum_{j=1}^{i-1}(\sigma_0 - j)\theta_j + \sum_{j=i}^{n-1}(\sigma_0 - j)\theta_{j+1} + c. \end{aligned} \tag{7.13}$$

Since for all $i \in N$,

$$\sigma_j(\theta_{N\setminus\{i\}}) = \begin{cases} j & \text{if } j < i, \\ j-1 & \text{if } j \geq i+1, \end{cases}$$

we can rewrite Eq. (7.13) as: for all $i \in N$,

$$g_i(\theta_{N\setminus\{i\}}) = \sum_{j \neq i}(\sigma_0 - \sigma_j(\theta_{N\setminus\{i\}}))\theta_j + c. \tag{7.14}$$

Observe that Eq. (7.14) holds no matter what the order of the efficient queue. This completes the proof. □

Our next theorem characterizes the complete family of rules satisfying *queue-efficiency*, *strategyproofness*, and *egalitarian equivalence*. It shows that the queue position in the reference bundle must be the same for all profiles. Furthermore, the choice of the queue position determines the transfers at all profiles up to a type-independent constant.

Theorem 7.1 *A rule μ satisfies queue-efficiency, strategyproofness, and egalitarian equivalence if and only if it is a VCG rule and there exists $\sigma_0 \in \{1, \ldots, n\}$ and $c \in \mathbf{R}$ such that for all $\theta \in \mathcal{Q}^N$ and all $i \in N$,*

$$t_i(\theta) = \sum_{j \in N\setminus\{i\}} (\sigma_0 - \sigma_j(\theta))\theta_j + c \tag{7.15}$$

where $\mu(\theta) = (\sigma(\theta), t(\theta))$.

[6]We can show Eq. (7.12) formally by induction. Note that Eq. (7.12) is true for $i = n-1$ by Eq. (7.11). The induction step—showing that Eq. (7.12) is true for $i-1$ if it is true for i—is easily established and is omitted.

Proof Lemma 7.1 has an immediate useful implication which we will use. By substituting Eq. (7.1) in Eq. (7.2) and simplifying, it follows that

$$t_0(\theta) = \sum_{i \in N} (\sigma_0 - \sigma_i(\theta))\theta_i + c. \tag{7.16}$$

We first show the necessity of Eq. (7.15). By *egalitarian equivalence,* $-(\sigma_i(\theta) - 1)\theta_i + t_i(\theta) = -(\sigma_0(\theta) - 1)\theta_i + t_0(\theta)$. Using Eq. (7.16) and Lemma 7.1, it follows that

$$t_i(\theta) = \sum_{j \neq i} (\sigma_0 - \sigma_j(\theta))\theta_j + c. \tag{7.17}$$

This establishes the necessity of Eq. (7.15). That σ_0 is any arbitrary queue position and c is an arbitrary constant follows from Lemma 7.1.

For sufficiency, consider a VCG mechanism μ such that the transfer satisfies Eq. (7.15), σ_0 is an arbitrary queue position, and c an arbitrary real constant. *Queue-efficiency* and *strategyproofness* are satisfied since μ is a VCG mechanism. We only need to show that it satisfies *egalitarian equivalence*. Let $i \in N$. Using Eq. (7.17), we obtain

$$\begin{aligned} u_i(\mu_i(\theta); \theta_i) &= -(\sigma_i(\theta) - 1)\theta_i - (\sigma_0 - \sigma_i(\theta))\theta_i + t_0(\theta) \\ &= -(\sigma_0 - 1)\theta_i + t_0(\theta) \\ &= u_i(\sigma_0, t_0(\theta); \theta_i). \end{aligned}$$

Since the selection of i is arbitrary, *egalitarian equivalence* follows. □

Remark 7.1 In this remark, we discuss what additional rules would be made possible by dropping one axiom at a time from the list appeared in Theorem 7.1.

(i) Dropping *queue-efficiency*: Let $\sigma_0 \in \{1, \ldots, n\}$ be any choice of queue position and $\sigma' \in \Sigma(N)$ be any queue. Let $\mu' = (\sigma, t')$ be a rule such that for all $\theta \in \mathcal{Q}^N$, $\sigma(\theta) = \sigma'$, and for all $i \in N$, $t'_i(\theta) = \sum_{j \in N \setminus \{i\}} (\sigma_0 - \sigma'_j)\theta_j$. Since $\sigma(\theta) = \sigma'$ for all θ, *queue-efficiency* is violated. It satisfies *strategyproofness* because, by misreporting, no agent can change her queue position or her own transfer. To show that *egalitarian equivalence* is satisfied, choose the reference bundle $(\sigma_0, t_0(\theta))$ where $t_0(\theta) = \sum_{j \in N} (\sigma_0 - \sigma'_j)\theta_j$. Then, for all $i \in N$,

$$\begin{aligned} -(\sigma'_i - 1)\theta_i + t'_i(\theta) &= -(\sigma'_i - 1)\theta_i + \sum_{j \in N \setminus \{i\}} (\sigma_0 - \sigma'_j)\theta_j \\ &= -(\sigma_0 - 1)\theta_i + \sum_{j \in N} (\sigma_0 - \sigma'_j)\theta_j. \end{aligned}$$

Therefore, μ' satisfies *egalitarian equivalence.*

(ii) Dropping *strategyproofness*: Let $\bar{\mu} = (\sigma, \bar{t})$ be a rule satisfying the following properties:

(1) For all $\theta \in \mathscr{Q}^N$, $\sigma(\theta) \in \mathit{Eff}(\theta)$, $\sigma_0(\theta) \in \{1, \ldots, n\}$, and for all $i \in N$, $\bar{t}_i(\theta) = \sum_{j \in N\setminus\{i\}}(\sigma_0(\theta) - \sigma_j(\theta))\theta_j$.

(2) There exist profiles $\theta, \theta' \in \mathscr{Q}^N$ such that $\sigma_0(\theta) \neq \sigma_0(\theta')$.

Due to (2) and Theorem 7.1, $\bar{\mu}$ must violate *queue-efficiency, strategyproofness*, or *egalitarian equivalence*. By (1), $\bar{\mu}$ satisfies *queue-efficiency*. To show that *egalitarian equivalence* is satisfied, choose the reference bundle for a profile as $(\sigma_0(\theta), t_0(\theta))$ where $t_0(\theta) = \sum_{j \in N}(\sigma_0(\theta) - \sigma_j(\theta))\theta_j$. Then, for all $i \in N$,

$$\begin{aligned} -(\sigma_i(\theta) - 1)\theta_i + \bar{t}_i(\theta) &= -(\sigma_i(\theta) - 1)\theta_i + \sum_{j \in N\setminus\{i\}} (\sigma_0(\theta) - \sigma_j(\theta))\theta_j \\ &= -(\sigma_0(\theta) - 1)\theta_i + \sum_{j \in N}(\sigma_0(\theta) - \sigma_j(\theta))\theta_j. \end{aligned}$$

The last expression is the utility from the reference bundle and shows that *egalitarian equivalence* is satisfied.[7] Hence $\bar{\mu}$ must violate *strategyproofness*.

(iii) Dropping *egalitarian equivalence*: By Holmström (1979), the family of all VCG rules become admissible. As shown in Theorem 7.1, not all VCG rules satisfy *egalitarian equivalence*.

7.3 Additionally Imposing Budget Balance or Feasibility

We now examine whether the rules characterized in Theorem 7.1 satisfy additional desirable properties. One such property is *budget balance* which requires that there be no net transfer into or out of the problem.

Budget balance: For all $N \in \mathscr{N}$ and all $\theta \in \mathscr{Q}^N$, $\sum_{i=1}^n t_i(\theta) = 0$.

A weakening of budget balance is *feasibility* which allows a rule to accumulate a budget surplus but not a deficit. So long as the accumulated surplus can be disposed of elsewhere, this can be justified. Otherwise, it is an efficiency loss.

It is worth noting that there are rules satisfying *queue-efficiency* and *strategyproofness* which run a budget surplus, an example of which is the well-known *pivotal rule*. This rule serves everyone in the nonincreasing order of their waiting costs. Each agent pays the sum of waiting costs of those served behind her. This rule runs a budget surplus at all profiles.

[7] As a matter of fact, our construction of transfers shows that any choice of queue position in the reference bundle can be made consistent with *egalitarian equivalence* provided we have unlimited freedom in selecting transfers.

Feasibility: For all $N \in \mathcal{N}$ and all $\theta \in \mathcal{Q}^N$, $\sum_{i=1}^n t_i(\theta) \leq 0$.

We begin with investigating the implications of additionally imposing *budget balance*. As it turns out, none of the rules characterized in Theorem 7.1 satisfies *budget balance*.

Proposition 7.1 *There is no rule satisfying queue-efficiency, strategyproofness, egalitarian equivalence, and budget balance.*

Proof From Theorem 7.1, it follows that a rule satisfies *queue-efficiency, strategyproofness,* and *egalitarian equivalence* only if the transfers t_i satisfy Eq. (7.15). Applying *budget balance* and then simplifying it, we obtain that for all $\theta \in \mathcal{Q}^N$,

$$\sum_{j \in N}(\sigma_j(\theta) - \sigma_0)\theta_j = \frac{nc}{n-1}. \tag{7.18}$$

We now have an impossibility since the left-hand side of Eq. (7.18) is dependent on θ (no matter how we choose σ_0), while the right-hand side is a constant. □

While *budget balanced* rules are not possible, it turns out—rather unexpectedly—that there are rules satisfying *feasibility* together with *queue-efficiency, strategyproofness,* and *egalitarian equivalence*. The following theorem characterizes all such rules. In particular, we show that $\sigma_0 = 1$ and $c \leq 0$.

Theorem 7.2 *A rule μ satisfies queue-efficiency, strategyproofness, egalitarian equivalence, and feasibility if and only if it is a VCG rule such that the transfers satisfy Eq. (7.15) with $\sigma_0 = 1$ and $c \leq 0$.*

Proof By Theorem 7.1, a rule satisfies *queue-efficiency, strategyproofness,* and *egalitarian equivalence* if and only if it is a VCG mechanism with the transfers given by Eq. (7.15). If $\sigma_0 = 1$ and $c \leq 0$, then for all $i \in N$, $t_i(\theta) = -\sum_{j \in N\setminus\{i\}}(\sigma_j(\theta) - 1)\theta_j + c \leq 0$ and hence *feasibility* holds. This establishes sufficiency.

To establish necessity, first note that given Eq. (7.15), a rule satisfies *queue-efficiency, strategyproofness, egalitarian equivalence,* and *feasibility* only if for each $\theta \in \mathcal{Q}^N$,

$$\sum_{i \in N} t_i(\theta) = (n-1)\sum_{i \in N}(\sigma_0 - \sigma_i(\theta))\theta_i + nc \leq 0. \tag{7.19}$$

Therefore, we only need to show that for Eq. (7.19) to be true, it is necessary that $c \leq 0$ and $\sigma_0 = 1$.

Suppose first that $c \leq 0$ is not true, that is, $c > 0$. Consider a profile θ such that for all $i \in N$, $\theta_i = 2c/n(n-1)$. Then,

$$\sum_{i \in N} t_i(\theta) = (n-1)\sum_{k=1}^{n} \frac{2c(\sigma_0 - k)}{n(n-1)} + nc = (2\sigma_0 - 1)c. \tag{7.20}$$

Observe that in Eq. (7.20), for any $\sigma_0 \in \{1, \ldots, n\}$, $(2\sigma_0 - 1)c > 0$ and we have a violation of Eq. (7.19). Therefore, $c \leq 0$.

Next, suppose that $\sigma_0 = 1$ is not true, that is, assume that $\sigma_0 = k$ where $k \in \{2, \ldots, n\}$. Consider a profile $\theta \in \mathcal{Q}^N$ such that $\theta_1 = \cdots = \theta_k = a > \theta_{k+1} = \cdots = \theta_n = b > 0$. For this profile,

$$\sum_{i \in N} t_i(\theta) = (n-1)\left[\sum_{r=1}^{k}(k-r)a - \sum_{r=k+1}^{n}(r-k)b\right] + nc$$
$$= (n-1)\left[\frac{k(k-1)a}{2} - \frac{(n-k)(n-k+1)b}{2}\right] + nc. \quad (7.21)$$

Since $k \in \{2, \ldots, n\}$ and $c \leq 0$, we can choose

$$a = \frac{1}{k(k-1)}\left[3 - \frac{2nc}{n-1}\right], \quad b = \frac{2}{n(n-1)}.$$

It is easy to verify that $a > b > 0$. Substituting these values in Eq. (7.21) and simplifying, we obtain

$$\sum_{i \in N} t_i(\theta) = (n-1)\left[\frac{3}{2} - \frac{(n-k)(n-k+1)}{n(n-1)}\right] > 0.$$

This violates Eq. (7.19) and proves that $\sigma_0 = 1$ is necessary. □

Remark 7.2 Let Γ be the family of rules that satisfies *queue-efficiency, strategyproofness, egalitarian equivalence,* and *feasibility*. From Theorem 7.2, it follows that if μ, $\mu' \in \Gamma$, then for each profile θ, the difference between the transfer of any agent $i \in N$ *across the two rules* μ *and* μ' is calculated by using the (agent and profile independent common) constant c. Since this constant c is restricted to be nonpositive, it follows that the rule $\mu \in \Gamma$ that *minimizes the budget surplus* is the one for which $c = 0$. It is also clear from Eq. (7.19) that even when $c = 0$, unlike Yengin (2012), one cannot place any upper bound on the budget surplus.

We conclude this section with a technical remark.

Remark 7.3 If we replace the *feasibility* requirement by the *no surplus condition* (that is, for all $\theta \in \mathcal{Q}^N$, $\sum_{i \in N} t_i(\theta) \geq 0$), then we can show the following result. A rule μ satisfies *queue-efficiency, strategyproofness, egalitarian equivalence,* and *no surplus condition* if and only if it is a VCG rule such that for all $\theta \in \mathcal{Q}^N$ and all $i \in N$, $t_i(\theta) = \sum_{j \in N \setminus \{i\}}(\sigma_0 - \sigma_j(\theta))\theta_j + c$, $\sigma_0 \geq (n+1)/2$ and $c \geq 0$. Interestingly, this is not exactly the dual of Theorem 7.2 since we have $\sigma_0 \geq (n+1)/2$. Moreover, it can also be shown that one cannot place any bound on the budget deficit.

7.4 Impossibility Results

Theorem 7.1 shows that there is a family of rules satisfying *queue-efficiency, strategyproofness,* and *egalitarian equivalence*. We now ask what happens if we impose two additional desirable properties: *weak group strategyproofness* and *no-envy*. Unfortunately, we obtain negative results.

7.4.1 Weak Group Strategyproofness and Egalitarian Equivalence

Weak group strategyproofness requires that there is no deviation by a group making all deviating members strictly better off.

Weak group strategyproofness: For all $N \in \mathcal{N}$, all $\theta, \theta' \in \mathcal{Q}^N$, and all $S \subset N$, if for all $j \in N \setminus S$, $\theta_j = \theta'_j$, then for at least one $i \in S$, $u_i(\mu_i(\theta); \theta_i) \geq u_i(\mu_i(\theta'); \theta_i)$.

This axiom has been used by Bogomolnaia and Moulin (2004), Moulin and Shenker (2001), Mutuswami (2005), Mitra and Mutuswami (2011), and others. It is obvious that *weak group strategyproofness* implies *strategyproofness*.[8]

Proposition 7.2 *For $n \geq 3$, there is no rule satisfying queue-efficiency, weak group strategyproofness, and egalitarian equivalence.*

Proof Let $N \in \mathcal{N}$ be such that $N = \{1, \ldots, n\}$ and $n \geq 3$, μ be a rule satisfying *queue-efficiency, weak group strategyproofness,* and *egalitarian equivalence*, and for all $\theta \in \mathcal{Q}^N$, $\mu(\theta) = (\sigma(\theta), t(\theta))$. Since *weak group strategyproofness* implies *strategyproofness,* we can use Theorem 7.1 to infer that the reference bundle for any $\theta \in \mathcal{Q}^N$ is $(\sigma_0, t_0(\theta))$ where $t_0(\theta) = \sum\limits_{j \in N} (\sigma_0 - \sigma_j(\theta))\theta_j + c$. The allocation of agent i is $(\sigma_i(\theta), t_i(\theta))$ where $t_i(\theta) = \sum\limits_{j \in N \setminus \{i\}} (\sigma_0 - \sigma_j(\theta))\theta_j + c$.

Consider profiles θ and θ' such that $\theta_1 > \cdots > \theta_n$ and, for all $i \in N$, $\theta'_i = \theta_i + x$, $x > 0$, $\theta'_i \in (\theta_i, \theta_{i-1})$ for all $i \in \{2, \ldots, n\}$. For these profiles, we show that there is a violation of *weak group strategyproofness* for all choices of σ_0.

1. $\sigma_0 = 1$: Let $(\theta'_2, \theta'_3, \theta_{N \setminus \{2,3\}})$ be the true profile. We can check that agents 2 and 3 can profitably manipulate via the misreports θ_2 and θ_3.
2. $\sigma_0 = n$: Let θ be the true profile. Here, agents $n-2$ and $n-1$ can profitably manipulate via the misreports θ'_{n-1} and θ'_{n-2}.

[8] One can formulate a stronger notion of *group strategyproofness* which requires that there does not exist a group deviation making all deviating members weakly better off and at least one member strictly better off. Unfortunately, in the context of queueing, this notion is incompatible with *queue-efficiency* as shown in Mitra and Mutuswami (2011).

3. $\sigma_0 \neq \{1, n\}$: Let $\left(\theta'_{\sigma_0+1}, \theta_{N\setminus\{\sigma_0+1\}}\right)$ be the true profile. In this case agents $\sigma_0 - 1$ and $\sigma_0 + 1$ can manipulate profitably via the misreports θ'_{σ_0-1} and θ_{σ_0+1}. □

Remark 7.4 When $n = 2$, all rules satisfying *queue-efficiency, strategyproofness,* and *egalitarian equivalence* are also *weakly group strategyproof.* In particular, putting $c = 0$ and $\sigma_0 = k$, $k = 1, 2$, gives us the k-pivotal mechanisms identified in Mitra and Mutuswami (2011) and shown to be *weakly group strategyproof.* Clearly, adding a constant to all transfers preserves the *weakly group strategyproof* property. It follows from the earlier discussion that putting $\sigma_0 = 1$ and $c \leq 0$ gives us *feasibility* as well in this case.

7.4.2 Strategyproofness, Egalitarian Equivalence, and No-Envy

We begin with a comparison between our results obtained using egalitarian equivalence and the results using *no-envy* (Foley 1967), which requires that no agent should end up with a higher utility by consuming what any other agent consumes (see Chap. 5 for details). The idea of this comparison is to see the differing implications of the two equity criteria in the context of queueing problems.

In contrast to *no-envy,* which implies *queue-efficiency, egalitarian equivalence* imposes no restriction on the choice of the queue provided we have freedom in choosing transfers. To see this, let $\theta \in \mathcal{Q}^N$ and $(\sigma_0(\theta), t_0(\theta))$ be the reference bundle. Suppose the rule chooses the queue $\sigma(\theta)$. It is easy to show that *egalitarian equivalence* requires the transfers to satisfy the following restrictions: for all $\theta \in \mathcal{Q}^N$ and all $i \in N$,

$$t_i(\theta) = (\sigma_i(\theta) - \sigma_0(\theta))\theta_i + t_0(\theta).$$

As a matter of fact, the difference between *no-envy* and *egalitarian equivalence* extends beyond their implications for *queue-efficiency*. As shown in Proposition 5.4, these two concepts are incompatible when there are at least four agents. No additional assumption is needed for this result. On the other hand, as discussed in Sect. 5.5, when there are two or three agents, *no-envy* and *egalitarian equivalence* are compatible even if *budget balance* is additionally required. We note that the incompatibility of *budget balance, no-envy,* and *egalitarian equivalence* for the problem of assignment of objects in a general quasi-linear framework with three agents was established by Thomson (1990).

Here, we ask another question: Are *no-envy, egalitarian equivalence,* and *strategyproofness* compatible when $n \leq 3$? The answer is yes. When $n = 2$, the k-pivotal mechanisms are an example. We have already observed that when there are just two agents, these mechanisms satisfy *queue-efficiency, strategyproofness,* and *egalitarian equivalence*. From the discussion in Sect. 6.3.2, they also satisfy *no-envy*.

When $n = 3$, the proof of Proposition 5.4 shows that we must have $\sigma_0 = 2$. Using Eq. (7.15), we can compute the transfers easily.[9] Assuming that the constant is zero, the transfers for a profile θ such that $\theta_1 > \theta_2 > \theta_3$ are

$$t_1(\theta) = -\theta_3,\ t_2(\theta) = \theta_1 - \theta_3,\ t_3 = \theta_1.$$

The corresponding utilities are

$$u_1(\theta) = -\theta_3,\ u_2(\theta) = \theta_1 - \theta_2 - \theta_3,\ u_3(\theta) = \theta_1 - 2\theta_3.$$

It is straightforward to check that no agent envies any other agent. Not surprisingly, this rule does not satisfy *budget balance*.

In a heterogenous object assignment problem with the option that each agent may be assigned more than one object or the null object, Yengin (2011) shows that *assignment efficiency,*[10] *strategyproofness, egalitarian equivalence,* and *no-envy* are compatible. Hence, the fact that in the queueing problem each agent is assigned only one queue position (one object) along with *queue-efficiency* imposes significant structure to drive this incompatibility in comparison to the general model of assigning objects.[11]

7.5 Conclusions

We show that the family of rules satisfying *queue-efficiency, strategyproofness,* and *egalitarian equivalence* is nonempty and provides a complete characterization of this family. However, none of these rules satisfies additional desirable properties such as *budget balance* or immunity to group deviations. This is hardly surprising.

From a more general mechanism design perspective, asking for *egalitarian equivalent* VCG rules to satisfy *budget balance* or *weak group strategyproofness* is indeed asking for too much. In particular, if we want *egalitarian equivalence,* then *strategyproofness* reduces the degrees of freedom of the reference bundle substantially by making the queue position fixed for all profiles. Therefore, one should not be disheartened to find a negative result with such strong properties. One rather surprising result is that there are *egalitarian equivalent* VCG rules that satisfy *feasibility*.

[9] Since *no-envy* implies *queue-efficiency,* the queue is determined to satisfy *queue-efficiency*.

[10] *Assignment efficiency* requires that there should be no feasible allocation at which all agents are weakly better off and at least one agent is strictly better off.

[11] The possibility of assigning the null object is important for the possibility result. Indeed, Yengin (2011) chooses the combination of the null object and money as the reference bundle.

References

Bogomolnaia, A., & Moulin, H. (2004). Random matching under dichotomous preferences. *Econometrica, 72*, 257–279.

Chun, Y., Mitra, M., & Mutuswami, S. (2014). Egalitarian equivalence and strategyproofness in the queueing problem. *Economic Theory, 56*, 425–442.

Clarke, E. H. (1971). Multi-part pricing of public goods. *Public Choice, 11*, 17–33.

Demange, G. (1984). Implementing efficient egalitarian equivalent allocations. *Econometrica, 52*, 1167–1177.

Dutta, B., & Vohra, R. (1993). A characterization of egalitarian equivalence. *Economic Theory, 3*, 465–479.

Foley, D. (1967). Resource allocation and the public sector. *Yale Economic Essays, 7*, 45–98.

Groves, T. (1973). Incentives in teams. *Econometrica, 41*, 617–631.

Holmström. B. (1979). Groves' schemes on restricted domains. *Econometrica, 47*, 1137–1144.

Mitra, M., & Mutuswami, S. (2011). Group strategyproofness in queueing models. *Games and Economic Behavior, 72*, 242–254.

Moulin, H., & Shenker, S. (2001). Strategyproof sharing of submodular costs: Budget balance versus efficiency. *Economic Theory, 18*, 511–533.

Mutuswami, S. (2005). Strategyproofness, non-bossiness and group strategyproofness in a cost sharing model. *Economics Letters, 89*, 83–88.

Pazner, E., & Schmeidler, D. (1978). Egalitarian equivalent allocations: A new concept of equity. *Quarterly Journal of Economics, 92*, 671–687.

Thomson, W. (1990). On the non-existence of envy-free and egalitarian-equivalent allocations in economies with indivisibilities. *Economics Letters, 34*, 227–229.

Vickrey, W. (1961). Counterspeculation, auctions and competitive sealed tenders. *Journal of Finance, 16*, 8–37.

Yengin, D. (2011). Axiomatizing political philosophy of distributive justice: Equivalence of no-envy and egalitarian-equivalence with welfare-egalitarianism. Research Paper No. 2011–19, School of Economics, University of Adelaide.

Yengin, D. (2012). Egalitarian-equivalent Groves mechanisms in the allocation of heterogenous objects. *Social Choice and Welfare, 38*, 137–160.

Chapter 8
Subgroup Additivity

8.1 Introduction

Subgroup additivity requires that a rule assigns the same expected relative utility to each agent whether an agent's expected relative utility is calculated from the problem involving all agents or from its subproblems with smaller numbers of agents where the subproblems are formed from the original problem. The relative utility of an agent from a rule is the utility of the agent in a given problem relative to her identical expected utility (that is, the utility that she expects to derive when all others have the same waiting cost as her and each of the queue position is assigned with equal probability). In this chapter, we investigate its implications for the queueing problem. As a result, we present characterizations of five important rules in the queueing problem: the minimal transfer rule (Maniquet 2003), the maximal transfer rule (Chun 2006), the pivotal rule, the reward-based pivotal rule (Mitra and Mutuswami 2011), and the symmetrically balanced VCG rule (Kayi and Ramaekers 2010; Mitra 2001; Suijs 1996).

Our notion of *subgroup additivity* can be compared with *converse consistency,* which requires that if there is a feasible allocation with the property that for all proper subgroups of agents the rule chooses the restriction of the allocation to the subgroup for the associated reduced problem this subgroup faces, then the allocation should be chosen for the problem. *Converse consistency* is appealing from both practical and computational point of views as it tells us that to evaluate an allocation for some possibly large group, it is enough to do so at the two-agent level (Peleg 1986; Thomson 2004). We ask a related question: Is it possible to solve the n-agent problem by solving its subproblems with a smaller number of agents? Instead of introducing a reduced problem, we try to find an answer by solving subproblems directly obtained from the original problem. Also, we note that *subgroup additivity* is an important issue in other contexts such as poverty measures and has been studied extensively under the name of *decomposability* (Foster et al. 1984).

© Springer International Publishing Switzerland 2016
Y. Chun, *Fair Queueing*, Studies in Choice and Welfare,
DOI 10.1007/978-3-319-33771-5_8

We first show that, given any set of n agents, for each $s \in \{2, \ldots, n-2\}$, if a rule satisfies *queue-efficiency, Pareto indifference, equal treatment of equals,* and *s-subgroup additivity,* then it also satisfies $(s+1)$-*subgroup additivity* implying that *2-subgroup additivity* is the strongest requirement among the class of rules satisfying *queue-efficiency, Pareto indifference,* and *equal treatment of equals.* Using *2-subgroup additivity,* we provide characterizations of the minimal transfer rule, the maximal transfer rule, the pivotal rule, and the reward-based pivotal rule. Using *3-subgroup additivity,* we provide characterizations of the symmetrically balanced VCG rule.

Given the *subgroup additivity* axiom and some basic axioms like *queue-efficiency, Pareto indifference,* and *equal treatment of equals,* we impose additionally two types of axioms to characterize the five rules. The first type of axioms is strategic where we weaken the notion of *strategyproofness. Strategyproofness,* discussed in Chap. 6, requires that an agent should not have an incentive to misrepresent her waiting cost no matter what she believes other agents to be doing. The second type of axioms is equity based where we modify the notion of *egalitarian equivalence* (Pazner and Schmeidler 1978). *Egalitarian equivalence,* discussed in Chap. 7, requires that there should be a reference bundle for each problem such that each agent is indifferent between her bundle and the reference bundle. We modify *egalitarian equivalence* by imposing a restriction on the reference bundle and requiring the axiom to be satisfied only for two-agent problems. What comes out of our analysis is that each strategic axiom in our characterization results can be replaced by an appropriate equity axiom for the four rules that are 2-additive. This type of substitutability between strategic and equity axioms also works for the symmetrically balanced VCG rule.[1]

This chapter is organized as follows.[2] In Sect. 8.2, we briefly discuss the importance of the five rules in the context of queueing and provide the preliminaries useful for this chapter. In Sect. 8.3, we present our main axiom of *subgroup additivity* and show that all five rules satisfy some version of this property. In Sect. 8.4, we provide our characterization result of the four 2-subgroup additive rules using strategic properties, and in Sect. 8.5, using equity properties. In Sect. 8.6, we characterize the symmetrically balanced VCG rule using *3-subgroup additivity* along with either a strategic axiom or an equity axiom. In Sect. 8.7, we conclude our analysis.

---

[1]A similar analysis can be done for scheduling problems in which agents have the same waiting costs, but the processing time for jobs can differ. For studies on the scheduling problem, see, for example, Cres and Moulin (2001) and Moulin (2007).

[2]This chapter is based mainly on Chun and Mitra (2014). Excerpts from that article are reprinted with kind permission of Elsevier.

8.2 Some Properties of Five Important Rules

We characterize five rules formally defined in Chap. 2. The minimal transfer rule (Maniquet 2003) selects an efficient queue and transfers to each agent a half of her unit waiting cost multiplied by the number of her predecessors minus a half of the sum of the unit waiting cost of her followers. The maximal transfer rule (Chun 2006) selects an efficient queue and transfers to each agent a half of the sum of the unit waiting cost of her predecessors minus a half of her unit waiting cost multiplied by the number of her followers. These two rules are obtained by applying the Shapley value (Shapley 1953), one of the most widely discussed solution in cooperative games, to appropriately designed cooperative games of the queueing problem. For the minimal transfer rule, the worth of a coalition is defined to be the minimal waiting cost incurred by its members under the assumption that they are served before the non-coalitional members. For the maximal transfer rule, the members of the coalition are served after the non-coalitional members. Both the minimum and the maximum transfer rules satisfy *queue-efficiency, budget balance, Pareto indifference,* and *equal treatment of equals.*[3]

The next three rules satisfy *strategyproofness,* that is, an agent should not have an incentive to misrepresent her waiting cost no matter what she believes other agents to be doing. The symmetrically balanced VCG rule is introduced by Suijs (1996) and Mitra (2001) and later characterized in the context of queueing by Kayi and Ramaekers (2010, in press) and Chun et al. (in press). It is the only rule satisfying *queue-efficiency, budget balance, Pareto indifference, equal treatment of equals,* and *strategyproofness.*

The pivotal and the reward-based pivotal rules fail to satisfy *budget balance* but are nevertheless important primarily because these rules satisfy *weak group strategyproof,* which requires that any subgroup of agents cannot be made strictly better off by deviating. They belong to the family of k-pivotal rules characterized by Mitra and Mutuswami (2011) by imposing the axioms of *queue-efficiency, equal treatment of equals, pairwise strategyproofness,* and *weak linearity.*[4] While *weak linearity* is a technical requirement, *pairwise strategyproofness* requires that, as long as there is no further side payments across agents, there does not exist any pair of agents that can benefit by deviating from truth telling. Moreover, all k-pivotal rules are *weakly group strategyproof.* We leave out other k-pivotal rules simply because they fail to satisfy *subgroup additivity.*

[3]*Queue-efficiency* requires that the rule should choose queues which minimize the total waiting costs. *Budget balance* requires that the sum of all transfers should be equal to zero. *Pareto indifference* requires that if an allocation is chosen by a rule, then all other feasible allocations which assign the same utilities to each agent should be chosen by the rule. Finally, *equal treatment of equals* requires that two agents with the same waiting cost should end up with the same utilities.

[4]As in Chaps. 6 and 7, since Mitra and Mutuswami (2011) assume the existence of a tie-breaking rule, which selects an efficient queue whenever there is more than one such queue, they do not impose *Pareto indifference.*

We note once again that the reward-based pivotal rule does not satisfy *budget balance*. Moreover, it does not satisfy *no budget deficit* either. Therefore, in this chapter, we weaken the definition of *feasibility* and modify the axiom of *Pareto indifference* to accommodate the reward-based pivotal rule. An allocation is *feasible* if no two agents are assigned the same position.[5] For all $N \in \mathcal{N}$ and all $\theta \in \mathcal{Q}^N$, let $Z(\theta)$ be the set of all feasible allocations. *Pareto indifference,* which requires that if an allocation is chosen by a rule, then all other feasible allocations which assign the same utilities to each agent should be chosen by the rule, is modified as follows.

Pareto indifference: For all $N \in \mathcal{N}$, all $\theta \in \mathcal{Q}^N$, all $(\sigma, t) \in \varphi(\theta)$, and $(\sigma', t') \in Z(\theta)$, if $\sigma' \in \mathit{Eff}(\theta)$ and for all $i \in N$, $u_i(\sigma'_i, t'_i; \theta_i) = u_i(\sigma_i, t_i; \theta_i)$, then $(\sigma', t') \in \varphi(\theta)$.

In the next section, we introduce *subgroup additivity* and we also study certain properties of subgroup additive rules by restricting our analysis to the class of all rules satisfying the three basic axioms of *queue-efficiency, Pareto indifference,* and *equal treatment of equals*. This class is not that restrictive in this context since all five rules satisfy the three axioms.

Remark 8.1 One advantage of having a rule φ satisfying *queue-efficiency, Pareto indifference,* and *equal treatment of equals* is that we can, without loss of generality, select any queue-efficient allocation from the set of all feasible allocations in a given problem. In particular, suppose that for some $N \in \mathcal{N}$ and $\theta \in \mathcal{Q}^N$, there exists i, $j \in N$ such that $\theta_i = \theta_j$. If $\sigma \in \mathit{Eff}(\theta)$, then the queue σ' obtained by interchanging the queue positions of agents i and j, ceteris paribus, also belongs to $\mathit{Eff}(\theta)$. By *equal treatment of equals,* if $(\sigma, t) \in \varphi(\theta)$, then $u_i(\sigma_i, t_i; \theta_i) = u_j(\sigma_j, t_j; \theta_j)$. Consider the transfer t' obtained from t by interchanging the transfers of agents i and j, ceteris paribus. Given $\theta_i = \theta_j$, by *Pareto indifference,* $(\sigma', t') \in \varphi(\theta)$. Hence we have $u_i(\sigma_i, t_i; \theta_i) = u_i(\sigma_j, t_j; \theta_i)$ and $u_j(\sigma_i, t_i; \theta_j) = u_j(\sigma_j, t_j; \theta_j)$ implying that, given any problem, each agent receives the same utility at any queue-efficient allocation of the problem.

From Remark 8.1, for simplicity of exposition, we use the following notation. For any given rule φ satisfying the three basic axioms of *queue-efficiency, Pareto indifference,* and *equal treatment of equals,* any $\theta \in \mathcal{Q}^N$ and any $(\sigma, t) \in \varphi(\theta)$, $u_i(\sigma_i, t_i; \theta_i) \equiv U_i(\theta) = U_i(\theta_i, \theta_{N\setminus\{i\}})$.

Finally, when the requirement of some axiom is restricted to problems involving exactly s agents, we prefix s. For example, *s-equal treatment of equal* is a weakening of *equal treatment of equals* which requires a rule to satisfy *equal treatment of equals* for all problems with exactly s-agents.

[5]Note that there is no restriction on the budget.

8.3 Subgroup Additivity

In general, it is more difficult to solve a problem involving all agents than solving several problems with smaller numbers of agents, and hence it is always helpful to look for rules that are subgroup additive. Before going into the formal definition of *subgroup additivity,* we introduce some relevant definitions and concepts. For any nonempty coalition S of agents and any agent $k \in \{1, \ldots, |S|\}$, let $T(S; k) = \{S' \subseteq S||S'| = k\}$, where $T(S; k)$ is the collection of all possible subgroups of S of size k and $|T(S; k)| = \binom{|S|}{k}$. Given any agent $i \in N$ and any coalition $S \subseteq N \setminus \{i\}$ with $|S| = k$, let the *identical expected utility of an agent i* with waiting cost $\theta_i > 0$ be $IEU(\theta_i, k + 1) = \frac{k}{2}\theta_i$. In a problem with $k + 1$ agents that includes agent i, if all other agents had the same waiting cost as agent i, then the identical expected utility of agent i is the cost that she incurs when she gets each of the $k + 1$ queue positions with equal probability $\frac{1}{k+1}$. Define the *relative* utility of agent i in problem $\theta_{S\cup\{i\}} = (\theta_i, \theta_S)$ as

$$\overline{U}_i(\theta_i, \theta_S) = \frac{U_i(\theta_i, \theta_S)}{IEU(\theta_i, |S \cup \{i\}|)},$$

where $\overline{U}_i(\theta_i, \theta_S)$ is the ratio of the actual utility and the identical expected utility. For any N, any $i \in N$ and any $s \in \{2, \ldots, |N \setminus \{i\}|\}$,

$$E_i(\theta \mid s) = \frac{1}{|T(N \setminus \{i\}; s - 1)|} \sum_{S' \in T(N\setminus\{i\};s-1)} \overline{U}_i(\theta_{S'\cup\{i\}}).$$

Therefore, $E_i(\theta \mid s)$ is the expected relative utility of agent i from all possible allocations with subgroups of size s (which always includes agent i) where the probability of selection of each subgroup of size $s - 1$ from the set of $N \setminus \{i\}$ agents is equally likely. A rule φ is *s-subgroup additive* if for all $N \in \mathcal{N}$ such that $|N| > s$, all $\theta \in \mathcal{Q}^N$, and all $i \in N$,

$$E_i(\theta \mid |N|) = E_i(\theta \mid s). \tag{8.1}$$

Using the definition of relative utility, we can simplify each term in Eq. (8.1) to obtain

$$\begin{aligned} E_i(\theta \mid s) &= \frac{1}{\binom{|N|-1}{s-1}} \sum_{S' \in T(N\setminus\{i\};s-1)} \frac{U_i(\theta_i, \theta_{S'})}{-(s-1)\frac{\theta_i}{2}} \\ &= -\frac{1}{\frac{\theta_i}{2}(|N| - 1)\binom{|N|-2}{s-2}} \sum_{S' \in T(N\setminus\{i\};s-1)} U_i(\theta_i, \theta_{S'}) \end{aligned} \tag{8.2}$$

and

$$E_i(\theta \mid |N|) = \frac{1}{\binom{|N|-1}{|N|-1}} \frac{U_i(\theta)}{-(|N|-1)\frac{\theta_i}{2}} = -\frac{U_i(\theta)}{(|N|-1)\frac{\theta_i}{2}}. \tag{8.3}$$

Equating Eqs. (8.2) and (8.3) and then simplifying it, we obtain

$$U_i(\theta) = \frac{1}{\binom{|N|-2}{s-2}} \sum_{S' \in T(N \setminus \{i\};s-1)} U_i(\theta_i, \theta_{S'}). \tag{8.4}$$

Using Eq. (8.4), we can rewrite the definition of *s-subgroup additivity* as follows.

***s*-subgroup additivity:** Given any positive integer s, for all $N \in \mathcal{N}$ with $|N| > s$, all $\theta \in \mathcal{Q}^N$, and all $i \in N$,

$$U_i(\theta) = \frac{1}{\binom{|N|-2}{s-2}} \sum_{S \subseteq N \setminus \{i\}, |S|=s-1} U_i(\theta_i, \theta_S). \tag{8.5}$$

Remark 8.2 One can make a comparison between *subgroup additivity* and *converse consistency*. Suppose that we want to assign utilities to each agent by solving $s(< n)$-agent problems instead of solving an n-agent problem. If a rule satisfies *s-subgroup additivity,* then each agent ends up with the same utility whether an n-agent problem is solved or whether we take an appropriately weighted sum of all the utilities assigned to this agent in all s-agent problems that includes her. As in the case of *converse consistency, subgroup additivity* provides a way of finding an allocation for the n-agent problem by focusing on problems with a smaller number of agents. However, the difference between *converse consistency* and *subgroup additivity* comes from the difference in the initial situation and in the nature of aggregation. For *converse consistency*, we start with the n-agent problem, and we look at the desirability of a feasible alternative simply by looking at the restriction on s-agent problem that results when all but these s agents have left with their components. This is different from *subgroup additivity* where the starting point is the s-agent problems and the weighted additivity aspect takes us to the n-agent problem.

We first investigate the relation between *subgroup additivity* axioms with different sizes of subgroups.

Proposition 8.1 *Let φ be a rule satisfying queue-efficiency, Pareto indifference, and s-equal treatment of equals. If the rule φ satisfies s-subgroup additivity with $s \in \{2, \ldots, |N|-2\}$, then it also satisfies $(s+1)$-subgroup additivity.*

Proof If a rule φ satisfies *queue-efficiency, Pareto indifference, s-equal treatment of equals,* and *s-subgroup additivity,* then for any given $N \in \mathcal{N}$ with $|N| \equiv n > s$, all $\theta \in \mathcal{Q}^N$, and all $i \in N$,

$$U_i(\theta) = \frac{1}{\binom{n-2}{s-2}} \sum_{S \subset N \setminus \{i\}, |S|=s-1} U_i(\theta_i, \theta_S).$$

If $s \in \{2, \dots, n-2\}$, then we have

$$\begin{aligned}
U_i(\theta) &= \frac{1}{\binom{n-2}{s-2}} \sum_{S \subset N\setminus\{i\}, |S|=s-1} U_i(\theta_i, \theta_S) \\
&= \frac{1}{\binom{n-2}{s-1}} \sum_{S \subset N\setminus\{i\}, |S|=s-1} \left\{ \binom{n-s}{s-1} U_i(\theta_i, \theta_S) \right\} \\
&= \frac{1}{\binom{n-2}{s-1}} \sum_{S' \subset N\setminus\{i\}, |S'|=s} \sum_{j \in S'} \frac{U_i(\theta_i, \theta_{S'\setminus\{j\}})}{s-1} \\
&= \frac{1}{\binom{n-2}{s-1}} \sum_{S' \subset N\setminus\{i\}, |S'|=s} \sum_{j \in S'} \frac{-\left(\sigma_i\left(\theta_i, \theta_{S'\setminus\{j\}}\right) - 1\right)\theta_i + t_i\left(\theta_i, \theta_{S'\setminus\{j\}}\right)}{s-1} \\
&= \frac{1}{\binom{n-2}{s-1}} \sum_{S' \subset N\setminus\{i\}, |S'|=s} \left\{ -\left(\sigma_i\left(\theta_i, \theta_{S'}\right) - 1\right)\theta_i + \sum_{j \in S'} \frac{t_i\left(\theta_i, \theta_{S'\setminus\{j\}}\right)}{s-1} \right\} \\
&= \frac{1}{\binom{n-2}{s-1}} \sum_{S' \subset N\setminus\{i\}, |S'|=s} U_i(\theta_i, \theta_{S'}).
\end{aligned}$$

Therefore, $(s+1)$-*subgroup additivity* is also satisfied. The last step follows by defining the transfer $t_i(\theta_i, \theta_{S'}) \equiv \frac{1}{s-1} \sum_{j \in S'} t_i\left(\theta_i, \theta_{S'\setminus\{j\}}\right)$ for each $S' \subset N \setminus \{i\}$ with $|S'| = s$. Finally, the penultimate step follows by noting that for any $S' \subset N \setminus \{i\}$ with $|S'| = s$,

$$\begin{aligned}
&\sum_{j \in S'} \frac{\left(\sigma_i\left(\theta_i, \theta_{S'\setminus\{j\}}\right) - 1\right)\theta_i}{s-1} \\
&= \sum_{j \in P_i(\sigma(\theta_i, \theta_{S'}))} \frac{\left(\sigma_i\left(\theta_i, \theta_{S'}\right) - 2\right)\theta_i}{s-1} + \sum_{j \in F_i(\sigma(\theta_i, \theta_{S'}))} \frac{\left(\sigma_i\left(\theta_i, \theta_{S'}\right) - 1\right)\theta_i}{s-1} \\
&= \left(\sigma_i\left(\theta_i, \theta_{S'}\right) - 1\right) \frac{\left(\sigma_i\left(\theta_i, \theta_{S'}\right) - 2\right)\theta_i}{s-1} + \left(s - \sigma_i\left(\theta_i, \theta_{S'}\right) + 1\right) \frac{\left(\sigma_i\left(\theta_i, \theta_{S'}\right) - 1\right)\theta_i}{s-1} \\
&= \left(\sigma_i\left(\theta_i, \theta_{S'}\right) - 1\right) \frac{\left(\sigma_i\left(\theta_i, \theta_{S'}\right) - 2\right)\theta_i}{s-1} \\
&\quad + \left\{(s-1) - \left(\sigma_i\left(\theta_i, \theta_{S'}\right) - 2\right)\right\} \frac{\left(\sigma_i\left(\theta_i, \theta_{S'}\right) - 1\right)\theta_i}{s-1} \\
&= \left(\sigma_i\left(\theta_i, \theta_{S'}\right) - 1\right)\theta_i.
\end{aligned}$$

□

Proposition 8.1 ensures that if a rule satisfies *queue-efficiency, Pareto indifference,* and *s-equal treatment of equals,* then *2-subgroup additivity* is the strongest, which implies *3-subgroup additivity,* and so on. The next proposition shows that if a rule satisfies *queue-efficiency, Pareto indifference,* and *s-equal treatment of equals,* then it is *s-subgroup additive* if and only if its transfer is *s-subgroup additive.*

Proposition 8.2 *Let φ be a rule satisfying queue-efficiency, Pareto indifference, and s-equal treatment of equals. The rule is s-subgroup additive if and only if the transfers are s-subgroup additive, that is, for all $N \in \mathcal{N}$ such that $|N| > s$, all $\theta \in \mathcal{Q}^N$, and all $i \in N$,*

$$t_i(\theta) = \frac{1}{\binom{|N|-2}{s-2}} \sum_{S \subset N \setminus \{i\}, |S|=s-1} t_i(\theta_i, \theta_S). \tag{8.6}$$

Proof If a rule φ is *s-subgroup additive,* then for all $N \in \mathcal{N}$ such that $|N| \equiv n > s$, all $\theta \in \mathcal{Q}^N$, and all $i \in N$,

$$U_i(\theta_i, \theta_{N \setminus \{i\}}) = \frac{1}{\binom{n-2}{s-2}} \sum_{S \subset N \setminus \{i\}, |S|=s-1} U_i(\theta_i, \theta_S). \tag{8.7}$$

Since $U_i(\theta_i, \theta_{N\setminus\{i\}}) = -(\sigma_i(\theta) - 1)\,\theta_i + t_i(\theta)$ and $U_i(\theta_i, \theta_S) = (\sigma_i(\theta_i, \theta_S) - 1)\,\theta_i + t_i(\theta_i, \theta_S)$, Eq. (8.7) can be rewritten as

$$-(\sigma_i(\theta) - 1)\,\theta_i + t_i(\theta) = \frac{-1}{\binom{n-2}{s-2}} \sum_{S \subset N \setminus \{i\}, |S|=s-1} \{(\sigma_i(\theta_i, \theta_S) - 1)\,\theta_i + t_i(\theta_i, \theta_S)\},$$

or equivalently,

$$t_i(\theta) - \frac{1}{\binom{n-2}{s-2}} \sum_{S \subset N \setminus \{i\}, |S|=s-1} t_i(\theta_i, \theta_S)$$
$$= (\sigma_i(\theta) - 1)\,\theta_i - \frac{1}{\binom{n-2}{s-2}} \sum_{S \subset N \setminus \{i\}, |S|=s-1} (\sigma_i(\theta_i, \theta_S) - 1)\,\theta_i.$$

Given the last step, it is enough to show that if the rule φ satisfies *queue-efficiency, Pareto indifference, s-equal treatment of equals,* and *s-subgroup additivity,* then

$$(A) \qquad (\sigma_i(\theta) - 1)\theta_i = \frac{1}{\binom{n-2}{n-2}} \sum_{S \subset N \setminus \{i\}, |S|=s-1} (\sigma_i(\theta_i, \theta_S) - 1)\theta_i.$$

If $\sigma_i(\theta) = 1$, then by *queue-efficiency, Pareto indifference,* and *s-equal treatment of equals,* we can find an allocation under the rule such that $P_i(\sigma(\theta)) = P_i(\sigma(\theta_i, \theta_S)) = \emptyset$ for all $S \subseteq N \setminus \{i\}$ and we have $\sigma_i(\theta_i, \theta_S) = 1$ for all $S \subseteq N \setminus \{i\}$

implying that condition (A) holds. If $|P_i(\sigma(\theta))| = \sigma_i(\theta) - 1 \in \{1, \ldots, n-1\}$, then

$$\begin{aligned}
&\frac{1}{\binom{n-2}{s-2}} \sum_{S \subset N\setminus\{i\}, |S|=s-1} (\sigma_i(\theta_i, \theta_S) - 1)\theta_i \\
&= \frac{1}{\binom{n-2}{s-2}} \left\{ \sum_{r=\max\{0, s-1-(n-\sigma_i(\theta))\}}^{\sigma_i(\theta)-1} r \binom{\sigma_i(\theta)-1}{r} \binom{n-\sigma_i(\theta)}{s-1-r} \right\} \theta_i \\
&= \frac{1}{\binom{n-2}{s-2}} (\sigma_i(\theta) - 1) \left\{ \sum_{r=\max\{0, s-1-(n-\sigma_i(\theta))\}+1}^{\sigma_i(\theta)-1} \binom{\sigma_i(\theta)-2}{r-1} \binom{n-\sigma_i(\theta)}{s-1-r} \right\} \theta_i \\
&= \frac{1}{\binom{n-2}{s-2}} (\sigma_i(\theta) - 1) \left\{ \binom{n-2}{s-2} \theta_i \right\} \\
&= (\sigma_i(\theta) - 1)\theta_i.
\end{aligned}$$

Therefore, condition (A) holds. □

Now we show that the five rules satisfy some version of *subgroup additivity*. The minimal transfer rule, the maximal transfer rule, the pivotal rule, and the reward-based pivotal rule satisfy *2-subgroup additivity*, whereas the symmetrically balanced rule satisfies *3-subgroup additivity*.

Proposition 8.3 *The minimal transfer rule, the maximal transfer rule, the pivotal rule, and the reward-based pivotal rule are all 2-subgroup additive. On the other hand, the symmetrically balanced rule is 3-subgroup additive.*

Proof It is easy to show that all five rules satisfy *queue-efficiency, Pareto indifference,* and *equal treatment of equals*. Hence to verify *subgroup additivity* of these five rules, all we need do is to check the restriction on transfer given in Eq. (8.6) of Proposition 8.2. In particular, to verify *2-subgroup additivity* of the four rules, we need to check whether for all $N \in \mathcal{N}$ such that $|N| > 2$, all $\theta \in \mathcal{Q}^N$, and all $i \in N$, $t_i(\theta) = \sum_{j \in N\setminus\{i\}} t_i(\theta_i, \theta_j)$. To prove *2-subgroup additivity,* we assume, without loss of generality, that if $j \in P_i(\sigma(\theta))$, then $j \in P_i(\sigma(\theta_i, \theta_j))$, and if $j \in F_i(\sigma(\theta))$, then $j \in F_i(\sigma(\theta_i, \theta_j))$ (see Remark 8.1). Similarly, to verify *3-subgroup additivity* of the symmetrically balanced VCG rule, we need to check whether for all $N \in \mathcal{N}$ such that $|N| > 3$, all $\theta \in \mathcal{Q}^N$, and all $i \in N$, $t_i(\theta) = \frac{1}{|N|-2} \sum_{\{j,k\} \subset N\setminus\{i\}} t_i(\theta_i, \theta_{\{j,k\}})$. Again to prove *3-subgroup additivity,* we assume, without loss of generality, that if $j \in P_i(\sigma(\theta))$, then $j \in P_i(\sigma(\theta_i, \theta_j, \theta_k))$ for all $k \in N \setminus \{i,j\}$, and if $j \in F_i(\sigma(\theta))$, then $j \in F_i(\sigma(\theta_i, \theta_j, \theta_k))$ for all $k \in N \setminus \{i,j\}$ (see Remark 8.1).

First, we begin with the minimum transfer rule φ^M. Let $N \in \mathcal{N}$ be such that $N = \{i,j\}$. If $\sigma_i(\theta_i, \theta_j) = 1$, then $t_i^M(\theta_i, \theta_j) = -\frac{\theta_j}{2}$ and $t_j^M(\theta_j, \theta_i) = \frac{\theta_j}{2}$. Therefore, for all $N \in \mathcal{N}$ such that $|N| > 2$, all $\theta \in \mathcal{Q}^N$, and all $i \in N$, it follows that

$$\begin{aligned} t_i^M(\theta) &= \frac{(\sigma_i(\theta)-1)\theta_i}{2} - \sum_{j \in F_i(\sigma(\theta))} \frac{\theta_j}{2} \\ &= \sum_{j \in P_i(\sigma(\theta))} t_i^M(\theta_i,\theta_j) + \sum_{j \in F_i(\sigma(\theta))} t_i^M(\theta_i,\theta_j) \\ &= \sum_{j \in N\setminus\{i\}} t_i^M(\theta_i,\theta_j). \end{aligned}$$

Hence φ^M satisfies *2-subgroup additivity*.

Next is the maximum transfer rule φ^C. Let $N \in \mathcal{N}$ be such that $N = \{i,j\}$. If $\sigma_i(\theta_i,\theta_j) = 1$, then $t_i^C(\theta_i,\theta_j) = -\frac{\theta_i}{2}$ and $t_j^C(\theta_j,\theta_i) = \frac{\theta_i}{2}$. Therefore, for all $N \in \mathcal{N}$ such that $|N| > 2$, all $\theta \in \mathcal{Q}^N$, and all $i \in N$, it follows that

$$\begin{aligned} t_i^C(\theta) &= \sum_{j \in P_i(\sigma(\theta))} \frac{\theta_j}{2} - \frac{(|N|-\sigma_i(\theta))\theta_i}{2} \\ &= \sum_{j \in P_i(\sigma(\theta))} t_i^C(\theta_i,\theta_j) + \sum_{j \in F_i(\sigma(\theta))} t_i^C(\theta_i,\theta_j) \\ &= \sum_{j \in N\setminus\{i\}} t_i^C(\theta_i,\theta_j). \end{aligned}$$

Hence φ^C satisfies *2-subgroup additivity*.

Now we consider the pivotal rule φ^P. Let $N \in \mathcal{N}$ be such that $N = \{i,j\}$. If $\sigma_i(\theta_i,\theta_j) = 1$, then $t_i^P(\theta_i,\theta_j) = -\theta_j$, and if $\sigma_i(\theta_i,\theta_j) = 2$, then $t_i^P(\theta_i,\theta_j) = 0$. Therefore, for all $N \in \mathcal{N}$ such that $|N| > 2$, all $\theta \in \mathcal{Q}^N$, and all $i \in N$, it can be shown that

$$\begin{aligned} t_i^P(\theta) &= -\sum_{j \in F_i(\sigma(\theta))} \theta_j \\ &= \sum_{j \in P_i(\sigma(\theta))} t_i^P(\theta_i,\theta_j) + \sum_{j \in F_i(\sigma(\theta))} t_i^P(\theta_i,\theta_j) \\ &= \sum_{j \in N\setminus\{i\}} t_i^P(\theta_i,\theta_j). \end{aligned}$$

Hence φ^P satisfies *2-subgroup additivity*.

Next is the reward-based pivotal rule φ^R. Let $N \in \mathcal{N}$ be such that $N = \{i,j\}$. If $\sigma_i(\theta_i,\theta_j) = 1$, then $t_i^R(\theta_i,\theta_j) = 0$, and if $\sigma_i(\theta_i,\theta_j) = 2$, then $t_i^R(\theta_i,\theta_j) = \theta_j$. Therefore, for all $N \in \mathcal{N}$ such that $|N| > 2$, all $\theta \in \mathcal{Q}^N$, and all $i \in N$, it can be shown that

$$\begin{aligned} t_i^R(\theta) &= \sum_{j \in P_i(\sigma(\theta))} \theta_j \\ &= \sum_{j \in P_i(\sigma(\theta))} t_i^R(\theta_i, \theta_j) + \sum_{j \in F_i(\sigma(\theta))} t_i^R(\theta_i, \theta_j) \\ &= \sum_{j \in N \setminus \{i\}} t_i^R(\theta_i, \theta_j). \end{aligned}$$

Hence φ^R satisfies *2-subgroup additivity.*

Finally, we consider the symmetrically balanced VCG rule φ^B. To prove *3-subgroup additivity* of φ^B, we have to show that for all $N \in \mathcal{N}$ such that $|N| > 3$, all $\theta \in \mathcal{Q}^N$, and all $i \in N$,

$$t_i^B(\theta) = \frac{1}{|N|-2} \sum_{\{j,k\} \subset N \setminus \{i\}} t_i^B(\theta_i, \theta_{\{j,k\}}).$$

Consider any $N \in \mathcal{N}$ such that $|N| > 3$, any $\theta \in \mathcal{Q}^N$, any $i \in N$, and any $S = \{i, j, k\} \subset N$. From the definition of φ^B, we have

$$t_i^B\left(\theta_i, \theta_{\{j,k\}}\right) = \begin{cases} \min\{\theta_j, \theta_k\} \text{ if } \{j,k\} \subseteq P_i(\sigma(0)), \\ -\max\{\theta_j, \theta_k\} \text{ if } \{j,k\} \subseteq F_i(\sigma(\theta)), \\ 0 \qquad \text{otherwise.} \end{cases} \tag{8.8}$$

Consider any $\ell \in P_i(\sigma(\theta_N))$ and all $\{j,k\} \subseteq N \setminus \{i\}$ such that $\ell \in \{j,k\}$ and $t_i^B\left(\theta_i, \theta_{\{j,k\}}\right) = \theta_\ell$. Given Eq. (8.8), this can happen for those $\{j,k\}$ such that $\ell \in \{j,k\}$ and the other agent $\ell' \in \{j,k\}$ is a predecessor of agent ℓ, that is, $\ell' \in P_\ell(\sigma(\theta))$. The total number of cases such that $t_i^B\left(\theta_i, \theta_{\{j,k\}}\right) = \theta_\ell$ is equal to $\binom{|P_\ell(\sigma(\theta))|}{1} = \sigma_\ell(\theta) - 1$. Similarly, consider any $\ell \in F_i(\sigma(\theta))$ and all $\{j,k\} \subseteq N \setminus \{i\}$ such that $\ell \in \{j,k\}$ and $t_i\left(\theta_i, \theta_{\{j,k\}}\right) = -\theta_\ell$. Given Eq. (8.8), this can happen for those $\{j,k\}$ such that $\ell \in \{j,k\}$ and the other agent $\ell' \in \{j,k\}$ is a follower of agent ℓ, that is, $\ell' \in F_\ell(\sigma(\theta))$. The total number of cases for which $t_i^B\left(\theta_i, \theta_{\{j,k\}}\right) = -\theta_\ell$ is $\binom{|F_\ell(\sigma(\theta))|}{1} = |N| - \sigma_\ell(\theta)$. Combining all these observations, it follows that for all $N \in \mathcal{N}$, all $\theta \in \mathcal{Q}^N$, and all $i \in N$,

$$\begin{aligned} \sum_{\{j,k\} \subset N \setminus \{i\}} \frac{t_i^B\left(\theta_i, \theta_{\{j,k\}}\right)}{|N|-2} &= \sum_{\ell \in P_i(\sigma(\theta))} \frac{\sigma_\ell(\theta) - 1}{|N|-2} \theta_\ell - \sum_{\ell \in F_i(\sigma(\theta))} \frac{|N| - \sigma_\ell(\theta)}{|N|-2} \theta_\ell \\ &= t_i^B(\theta). \end{aligned}$$

Hence φ^B satisfies *3-subgroup additivity.* □

8.4 2-Subgroup Additivity and Strategic Considerations

Strategyproofness discussed in Chap. 6 requires that an agent cannot strictly gain by misrepresenting her waiting cost no matter what she believes other agents to be doing. Here we introduce two weakenings of *strategyproofness* by requiring that it is applied separately to agents with extreme waiting costs and characterize the four 2-subgroup additive rules. *Strategyproofness for the highest cost agent* requires that an upward misrepresentation of the waiting cost by the agent with the highest waiting cost should not affect her utility. On the other hand, *strategyproofness for the lowest cost agent* requires that a downward misrepresentation of the waiting cost by the agent with lowest waiting cost should not affect her utility.

Strategyproofness for the highest cost agent: For all $N \in \mathcal{N}$, all $\theta, \theta' \in \mathcal{Q}^N$, all $(\sigma, t) \in \varphi(\theta)$, all $(\sigma', t') \in \varphi(\theta')$, and all $i \in N$, if $\sigma_i = 1$, $\theta'_i > \theta_i$, and for all $j \in N \setminus \{i\}$, $\theta'_j = \theta_j$, then $u_i(\sigma_i, t_i; \theta_i) = u_i(\sigma'_i, t'_i; \theta_i)$.

Strategyproofness for the lowest cost agent: For all $N \in \mathcal{N}$, all $\theta, \theta' \in \mathcal{Q}^N$, all $(\sigma, t) \in \varphi(\theta)$, all $(\sigma', t') \in \varphi(\theta')$, and all $i \in N$, if $\sigma_i = |N|$, $\theta'_i < \theta_i$, and for all $j \in N \setminus \{i\}$, $\theta'_j = \theta_j$, then $u_i(\sigma_i, t_i; \theta_i) = u_i(\sigma'_i, t'_i; \theta_i)$.

The minimal and the maximal transfer rules satisfy *2-subgroup additivity,* but they are not *strategyproof.* However, it is interesting to note that the minimal transfer rule does not give an agent with the highest waiting cost an incentive to misrepresent her true waiting cost upward and the maximal transfer rule does not give an agent with the lowest waiting cost an incentive to misrepresent her true waiting cost downward.

Next we ask whether it is possible to find rules that satisfy both *2-subgroup additivity* along with *strategyproofness* or something stronger like *weak group strategyproofness*? If we are willing to sacrifice the requirement of *budget balance,* the answer is yes. The pivotal and the reward-based pivotal rules are such examples. Moreover, these two rules can be characterized by replacing *budget balance* with exactly one of the two following axioms.

Last-agent zero transfer: For all $N \in \mathcal{N}$, all $\theta \in \mathcal{Q}^N$, all $(\sigma, t) \in \varphi(\theta)$, and all $i \in N$, if $\sigma_i = |N|$, then $t_i = 0$.

First-agent zero transfer: For all $N \in \mathcal{N}$, all $\theta \in \mathcal{Q}^N$, all $(\sigma, t) \in \varphi(\theta)$, and all $i \in N$, if $\sigma_i = 1$, then $t_i = 0$.

Now we present our first characterization of the four 2-subgroup additive rules based on the strategic axioms.

Theorem 8.1

(i) *The minimal transfer rule φ^M is the only rule satisfying queue-efficiency, budget balance, Pareto indifference, 2-equal treatment of equals, strategyproofness for the highest cost agent, and 2-subgroup additivity.*

(ii) The maximal transfer rule φ^C is the only rule satisfying queue-efficiency, budget balance, Pareto indifference, 2-equal treatment of equals, strategyproofness for the lowest cost agent, and 2-subgroup additivity.

(iii) The pivotal rule φ^P is the only rule satisfying queue-efficiency, last-agent zero transfer, Pareto indifference, 2-equal treatment of equals, strategyproofness for the highest cost agent, and 2-subgroup additivity.

(iv) The reward-based pivotal rule φ^R is the only rule satisfying queue-efficiency, first-agent zero transfer, Pareto indifference, 2-equal treatment of equals, strategyproofness for the lowest cost agent, and 2-subgroup additivity.

Proof

(i) It is easy to show that the minimal transfer rule φ^M satisfies *queue-efficiency, budget balance, Pareto indifference, 2-equal treatment of equals, strategyproofness for the highest cost agent,* and *2-subgroup additivity*. To prove the converse statement, let φ be a rule satisfying the six axioms. We begin with a two-agent problem. To simplify the notation, we assume without loss of generality that $N \in \mathcal{N}$ and $\theta \in \mathcal{Q}^N$ be such that $N = \{1, 2\}$ and $\theta_1 \geq \theta_2$. Let $\theta' \in \mathcal{Q}^N$ be such that $\theta' = (\theta_2, \theta_2)$ and $(\sigma', t') \in \varphi(\theta')$. By *queue-efficiency* and *Pareto indifference,* we may assume that for all $i \in N$, $\sigma'_i = i$. By *budget balance* and *2-equal treatment of equals,* $t'_1 = -\frac{\theta_2}{2}$ and $t'_2 = \frac{\theta_2}{2}$. Now suppose that the first agent makes an upward misrepresentation of her waiting cost, so that the resulting problem becomes θ. Let $(\sigma, t) \in \varphi(\theta)$. By *queue-efficiency* and *Pareto indifference,* we may assume that for all $i \in N$, $\sigma_i = i$. By *strategyproofness for the highest cost agent,* $t_1 = -\frac{\theta_2}{2}$, and by *budget balance,* $t_2 = \frac{\theta_2}{2}$, which shows that on the class of two-agent problems, $\varphi = \varphi^M$. By *2-subgroup additivity,* it follows that $\varphi = \varphi^M$ for all problems.

(ii) It is easy to show that the maximal transfer rule φ^C satisfies *queue-efficiency, budget balance, Pareto indifference, 2-equal treatment of equals, strategyproofness for the lowest cost agent,* and *2-subgroup additivity*. To prove the converse statement, let φ be a rule satisfying the six axioms. We begin with a two-agent problem. To simplify the notation, we assume without loss of generality that $N \in \mathcal{N}$ and $\theta \in \mathcal{Q}^N$ be such that $N = \{1, 2\}$ and $\theta_1 \geq \theta_2$. Let $\theta' \in \mathcal{Q}^N$ be such that $\theta' = (\theta_1, \theta_1)$ and $(\sigma', t') \in \varphi(\theta')$. By *queue-efficiency* and *Pareto indifference,* we may assume that for all $i \in N$, $\sigma'_i = i$. By *budget balance* and *2-equal treatment of equals,* $t'_1 = -\frac{\theta_1}{2}$ and $t'_2 = \frac{\theta_1}{2}$. Now suppose that the last agent makes a downward misrepresentation of her waiting cost, so that the resulting problem becomes θ. Let $(\sigma, t) \in \varphi(\theta)$. By *queue-efficiency* and *Pareto indifference,* we may assume that for all $i \in N$, $\sigma_i = i$. By *strategyproofness for the lowest cost agent,* $t_2 = \frac{\theta_1}{2}$, and by *budget balance,* $t_1 = -\frac{\theta_1}{2}$, which shows that on the class of two-agent problems, $\varphi = \varphi^C$. By *2-subgroup additivity,* it follows that $\varphi = \varphi^C$ for all problems.

(iii) It is easy to show that the pivotal rule φ^P satisfies *queue-efficiency, last-agent zero transfer, Pareto indifference, 2-equal treatment of equals, strategyproof-*

ness for the highest cost agent, and *2-subgroup additivity*. To prove the converse statement, let φ be a rule satisfying the six axioms. We begin with a two-agent problem. To simplify the notation, we assume without loss of generality that $N \in \mathcal{N}$ and $\theta \in \mathcal{Q}^N$ be such that $N = \{1,2\}$ and $\theta_1 \geq \theta_2$. Let $\theta' \in \mathcal{Q}^N$ be such that $\theta' = (\theta_2, \theta_2)$ and $(\sigma', t') \in \varphi(\theta')$. By *queue-efficiency* and *Pareto indifference,* we may assume that for all $i \in N$, $\sigma'_i = i$. By *last-agent zero transfer* and *2-equal treatment of equals,* $t'_1 = -\theta_2$ and $t'_2 = 0$. Now suppose that the first agent makes an upward misrepresentation of her waiting cost, so that the resulting problem becomes θ. Let $(\sigma, t) \in \varphi(\theta)$. By *queue-efficiency* and *Pareto indifference,* we may assume that for all $i \in N$, $\sigma_i = i$. By *strategyproofness for the highest cost agent,* $t_1 = -\theta_2$ and by *last-agent zero transfer,* $t_2 = 0$, which shows that on the class of two-agent problems, $\varphi = \varphi^P$. By *2-subgroup additivity,* it follows that $\varphi = \varphi^P$ for all problems.

(iv) It is easy to show that the reward-based pivotal rule φ^R satisfies *queue-efficiency, first-agent zero transfer, Pareto indifference, 2-equal treatment of equals, strategyproofness for the lowest cost agent,* and *2-subgroup additivity*. To prove the converse statement, let φ be a rule satisfying the six axioms. We begin with a two-agent problem. To simplify the notation, we assume without loss of generality that $N \in \mathcal{N}$ and $\theta \in \mathcal{Q}^N$ be such that $N = \{1,2\}$ and $\theta_1 \geq \theta_2$. Let $\theta' \in \mathcal{Q}^N$ be such that $\theta' = (\theta_1, \theta_1)$ and $(\sigma', t') \in \varphi(\theta')$. By *queue-efficiency* and *Pareto indifference,* we may assume that for all $i \in N$, $\sigma'_i = i$. By *first-agent zero transfer* and *2-equal treatment of equals,* $t'_1 = 0$ and $t'_2 = \theta_1$. Now suppose that the last agent makes a downward misrepresentation of her waiting cost, so that the resulting problem becomes θ. Let $(\sigma, t) \in \varphi(\theta)$. By *queue-efficiency* and *Pareto indifference,* we may assume that for all $i \in N$, $\sigma_i = i$. By *strategyproofness for the lowest cost agent,* $t_2 = \theta_1$, and by *first-agent zero transfer,* $t_1 = 0$, which shows that on the class of two-agent problems, $\varphi = \varphi^R$. By *2-subgroup additivity,* it follows that $\varphi = \varphi^R$ for all problems. □

8.5 2-Subgroup Additivity and Equity Considerations

Egalitarian equivalence (Pazner and Schmeidler 1978) requires that there should be a reference bundle for each problem such that each agent is indifferent between her bundle and the reference bundle. As shown in Chap. 7, there is a restricted family of rules satisfying *queue-efficiency, strategyproofness,* and *egalitarian equivalence* together. Moreover, none of these rules satisfy *2-* or *3-subgroup additivity*. In this section, by requiring *egalitarian equivalence* to hold only for two-agent problems, we characterize the four 2-subgroup additive rules.

2-egalitarian equivalence: For all $N \in \mathcal{N}$ such that $|N| = 2$, all $\theta \in \mathcal{Q}^N$, all $(\sigma, t) \in \varphi(\theta)$, and all $i \in N$, there exists a queue position $\sigma_0(\theta)$ and a transfer $t_0(\theta)$ such that $u_i(\sigma_i, t_i; \theta_i) = u_i(\sigma_0(\theta), t_0(\theta); \theta_i)$.

It is obvious that *2-egalitarian equivalence* implies *2-equal treatment of equals.*

For our next result, we use two stronger versions of *2-egalitarian equivalence* where, in the first version, the queue position in the reference bundle is fixed for all two-agent problems and, in the second version, the queue position in the reference bundle is fixed for all two-agent problems, and it is additionally required that there exists at least one profile for the two-agent problem for which the reference transfer is nonzero.

2-egalitarian equivalence with fixed queue position: For all $N \in \mathcal{N}$ such that $|N| = 2$, all $\theta \in \mathcal{Q}^N$, all $(\sigma, t) \in \varphi(\theta)$, and all $i \in N$, there exists a queue position $\sigma_0 \in \{1, 2\}$ and a transfer $t_0(\theta) \in \mathbf{R}^N$ such that $u_i(\sigma_i, t_i; \theta_i) = u_i(\sigma_0, t_0(\theta); \theta_i)$.[6]

2-strong egalitarian equivalence with fixed queue position: For all $N \in \mathcal{N}$ such that $|N| = 2$, all $\theta \in \mathcal{Q}^N$, all $(\sigma, t) \in \varphi(\theta)$, and all $i \in N$, there exists a queue position $\sigma_0 \in \{1, 2\}$ and a transfer $t_0(\theta) \in \mathbf{R}^N$ such that $u_i(\sigma_i, t_i; \theta_i) = u_i(\sigma_0, t_0(\theta); \theta_i)$, and for at least one $\theta' \in \mathcal{Q}^N$, $t_0(\theta') \neq 0$.

It is obvious that *2-strong egalitarian equivalence with fixed queue position* implies *2-egalitarian equivalence with fixed queue position,* which in turn implies *2-egalitarian equivalence* and *2-equal treatment of equals.*

Now we are ready to present our second characterization of the four 2-subgroup additive rules based on the equity axioms.

Theorem 8.2

(i) A rule satisfies queue-efficiency, budget balance, Pareto indifference, 2-egalitarian equivalence with fixed queue position, and 2-subgroup additivity if and only if it is either the minimal transfer rule or the maximal transfer rule.

(ii) A rule satisfies queue-efficiency, last-agent zero transfer, Pareto indifference, 2-strong egalitarian equivalence with fixed queue position, and 2-subgroup additivity if and only if it is the pivotal rule.

(iii) A rule satisfies queue-efficiency, first-agent zero transfer, Pareto indifference, 2-strong egalitarian equivalence with fixed queue position, and 2-subgroup additivity if and only if it is the reward-based pivotal rule.

Proof

(i) It is easy to show that the minimal transfer rule φ^M and the maximal transfer rule φ^C satisfy *queue-efficiency, budget balance, Pareto indifference, 2-egalitarian equivalence with fixed queue position,* (hence, *2-equal treatment of equals,*) and *2-subgroup additivity*. To prove the converse statement, let φ

[6]Strictly speaking, we should write $t_0(\sigma(\theta))$ instead of $t_0(\theta)$ since t_0 depends on which σ is chosen for each θ. However, to simplify the notation, we use $t_0(\theta)$ when there is no danger of confusion.

be a rule satisfying the five axioms. We begin with a two-agent problem. To simplify the notation, we assume without loss of generality that $N \in \mathcal{N}$ and $\theta \in \mathcal{Q}^N$ be such that $N = \{1, 2\}$ and $\theta_1 \geq \theta_2$. Let $(\sigma, t) \in \varphi(\theta)$. By *queue-efficiency* and *Pareto indifference,* we may assume that for all $i \in N$, $\sigma_i = i$. By *budget balance* and *2-egalitarian equivalence with fixed queue position,* we can have only the following two possibilities:

(i.1) $\sigma_0(\theta) = 1$, $t_0(\theta) = -\frac{\theta_2}{2}$, and $t = t^M$.
(i.2) $\sigma_0(\theta) = 2$, $t_0(\theta) = \frac{\theta_1}{2}$, and $t = t^C$.

By *2-egalitarian equivalence with fixed queue position,* if (i.1) is chosen for $\theta \in \mathcal{Q}^N$, then (i.1) should be chosen for all problems $\theta' \in \mathcal{Q}^N$ with $|N| = 2$ and hence by *2-subgroup additivity,* we obtain $\varphi = \varphi^M$ for all problems. Similarly, from *2-egalitarian equivalence with fixed queue position,* if (i.2) is chosen for $\theta \in \mathcal{Q}^N$, then (i.2) should also be chosen for all problems $\theta' \in \mathcal{Q}^N$ with $|N| = 2$ and hence by *2-subgroup additivity,* we obtain $\varphi = \varphi^C$ for all problems. Hence if a rule φ satisfies the five axioms, then we conclude that $\varphi \in \{\varphi^M, \varphi^C\}$.

(ii) It is easy to show that the pivotal rule φ^P satisfies *queue-efficiency, last-agent zero transfer, Pareto indifference, 2-strong egalitarian equivalence with fixed queue position,* (hence, *2-equal treatment of equals,*) and *2-subgroup additivity*. To prove the converse statement, let φ be a rule satisfying the five axioms. We begin with a two-agent problem. To simplify the notation, we assume without loss of generality that $N \in \mathcal{N}$ and $\theta \in \mathcal{Q}^N$ be such that $N = \{1, 2\}$ and $\theta_1 \geq \theta_2$. Let $(\sigma, t) \in \varphi(\theta)$. By *queue-efficiency* and *Pareto indifference,* we may assume that for all $i \in N$, $\sigma_i = i$. By *last-agent zero transfer* and *2-strong egalitarian equivalence with fixed queue position,* we can have only the following two possibilities:

(ii.1) $\sigma_0(\theta) = 1$, $t_0(\theta) = -\theta_2$, $t_1 = -\theta_2$, and $t_2 = 0$. Therefore, $t = t^P$.
(ii.2) $\sigma_0(\theta) = 2$, $t_0(\theta) = 0$, $t_1 = -\theta_1$, and $t_2 = 0$.

Note that (ii.2) fails to satisfy *2-strong egalitarian equivalence with fixed queue position* since $\sigma_0(\theta) = 2$ and hence $t_0(\theta) = 0$ for all profiles. We are left with (ii.1) which satisfies *2-strong egalitarian equivalence with fixed queue position*. Therefore, if a rule φ satisfies the five axioms, then it coincides with the pivotal rule on the class of two-agent problems, and by *2-subgroup additivity,* we conclude that $\varphi = \varphi^P$ for all problems.

(iii) It is easy to show that the reward-based pivotal rule φ^R satisfies *queue-efficiency, first-agent zero transfer, Pareto indifference, 2-strong egalitarian equivalence with fixed queue position,* (hence, *2-equal treatment of equals*), and *2-subgroup additivity*. To prove the converse statement, let φ be a rule satisfying the five axioms. We begin with a two-agent problem. To simplify the notation, we assume without loss of generality that $N \in \mathcal{N}$ and $\theta \in \mathcal{Q}^N$ be such that $N = \{1, 2\}$ and $\theta_1 \geq \theta_2$. Let $(\sigma, t) \in \varphi(\theta)$. By *queue-efficiency* and *Pareto indifference,* we may assume that for all $i \in N$, $\sigma_i = i$. By *first-agent*

zero transfer and *2-strong egalitarian equivalence with fixed queue position*, we can have only the following two possibilities:

(iii.1) $\sigma_0(\theta) = 1$, $t_0(\theta) = 0$, $t_1 = 0$, and $t_2 = \theta_2$.
(iii.2) $\sigma_0(\theta) = 2$, $t_0(\theta) = \theta_1$, $t_1 = 0$, and $t_2 = \theta_1$. Therefore, $t = t^R$.

In this case, (iii.1) fails to satisfy *2-strong egalitarian equivalence with fixed queue position* since $\sigma_0(\theta) = 1$ and hence $t_0(\theta) = 0$ for all profiles, and (iii.2) satisfies *2-strong egalitarian equivalence with fixed queue position*. Therefore, if a rule φ satisfies the five axioms, then it coincides with the reward-based pivotal rule on the class of two-agent problems, and by *2-subgroup additivity*, we conclude that $\varphi = \varphi^R$ for all problems.

8.6 3-Subgroup Additivity and Strategic/Equity Considerations

Of the five important rules only the symmetrically balanced VCG rule fails to satisfy *2-subgroup additivity* since it is defined for all problems with more than two agents. However, as shown in Proposition 8.3, it satisfies *3-subgroup additivity*. In this section, we present two characterization results of the symmetrically balanced VCG rule. The first characterization imposes a strategic axiom and the second an equity axiom. *Strategyproofness for extreme cost agents* requires that either an upward manipulation of the waiting cost by the agent with the highest waiting cost or a downward manipulation by the agent with the lowest waiting cost should not affect the utility of the agent involved. Clearly, it is weaker than *strategyproofness* but stronger than both *strategyproofness for the highest cost agent* and *strategyproofness for the lowest cost agent.*

Strategyproofness for extreme cost agents: For all $N \in \mathcal{N}$, all $\theta, \theta' \in \mathcal{Q}^N$, all $(\sigma, t) \in \varphi(\theta)$, all $(\sigma', t') \in \varphi(\theta')$, and all $i \in N$, if either $[\sigma_i = 1, \theta'_i > \theta_i$, and for all $j \in N \setminus \{i\}]$ or $[\sigma_i = |N|, \theta'_i < \theta_i$, and for all $j \in N \setminus \{i\}, \theta'_j = \theta_j]$, then $u_i(\sigma_i, t_i; \theta_i) = u_i(\sigma'_i, t'_i; \theta_i)$.

Our second characterization of the symmetric balanced VCG rule uses the following axiom which requires that for all three-agent problems, there is a fixed queue position such that an agent in the fixed position is indifferent among all three bundles. It is an equity requirement saying that the allocation should not favor one agent at the expense of another from the perspective of at least one agent in the fixed position.

3-positional equivalence: For all $N \in \mathcal{N}$ such that $|N| = 3$, all $\theta \in \mathcal{Q}^N$, all $(\sigma, t) \in \varphi(\theta)$, and all $i \in N$, there exists a queue position $\sigma_0 \in \{1, 2, 3\}$ such that if $\sigma_i = \sigma_0$, then for all $j \in N$, $u_i(\sigma_i, t_i; \theta_i) = u_i(\sigma_j, t_j; \theta_i)$.

It is not difficult to show that the symmetrically balanced VCG rule satisfies *3-positional equivalence* since the agent in the middle position is indifferent among the three bundles, $(\sigma_i, t_i)_{i \in N}$, for all three-agent problems.

Now we are ready to present our two characterizations of the symmetrically balanced VCG rule.

Theorem 8.3

(i) The symmetrically balanced VCG rule is the only rule satisfying queue-efficiency, budget balance, Pareto indifference, 3-equal treatment of equals, strategyproofness for extreme cost agents, and 3-subgroup additivity.

(ii) The symmetrically balanced VCG rule is the only rule satisfying queue-efficiency, budget balance, Pareto indifference, 3-equal treatment of equals, 3-positional equivalence, and 3-subgroup additivity.

Proof

(i) It is obvious that the symmetrically balanced VCG rule φ^B satisfies *queue-efficiency, budget balance, Pareto indifference, 3-equal treatment of equals, strategyproofness for extreme cost agents,* and *3-subgroup additivity*. To prove the converse statement, let φ be a rule satisfying the six axioms. We begin with a three-agent problem. To simplify the notation, we assume without loss of generality that $N \in \mathcal{N}$ and $\theta \in \mathcal{Q}^N$ be such that $N = \{1, 2, 3\}$ and $\theta_1 \geq \theta_2 \geq \theta_3$. Let $\theta^1 \in \mathcal{Q}^N$ be such that $\theta^1 = (\theta_2, \theta_2, \theta_2)$ and $(\sigma^1, t^1) \in \varphi(\theta^1)$. By *queue-efficiency* and *Pareto indifference,* we may assume that for all $i \in N$, $\sigma_i^1 = i$. By *budget balance* and *3-equal treatment of equals,* $t_1^1 = -\theta_2$, $t_2^1 = 0$, and $t_3^1 = \theta_2$.

Suppose that the first agent makes an upward misrepresentation of her waiting cost from what it was in θ^1, so that the resulting problem becomes $\theta^2 \in \mathcal{Q}^N$ where $\theta^2 = (\theta_1, \theta_2, \theta_2)$. Let $(\sigma^2, t^2) \in \varphi(\theta^2)$. By *queue-efficiency* and *Pareto indifference,* we may assume that for all $i \in N$, $\sigma_i^2 = i$. By *strategyproofness for extreme cost agents,* $t_1^2 = -\theta_2$. By *budget balance* and *3-equal treatment of equals,* $t_2^2 = 0$ and $t_3^2 = \theta_2$ implying that $t^2 = t^1$.

Next suppose that the last agent makes a downward misrepresentation of her waiting cost from what it was in θ^1, so that the resulting problem becomes $\theta^3 \in \mathcal{Q}^N$ where $\theta^3 = (\theta_2, \theta_2, \theta_3)$. Let $(\sigma^3, t^3) \in \varphi(\theta^3)$. By *queue-efficiency* and *Pareto indifference,* we may assume that for all $i \in N$, $\sigma_i^3 = i$. By *strategyproofness for extreme cost agents,* $t_3^3 = \theta_2$. By *budget balance* and *3-equal treatment of equals,* $t_1^3 = -\theta_2$ and $t_2^3 = 0$ implying that $t^3 = t^1$.

Now, consider the initial problem $\theta \in \mathcal{Q}^N$ such that $\theta_1 \geq \theta_2 \geq \theta_3$. Let $(\sigma, t) \in \varphi(\theta)$. By *queue-efficiency* and *Pareto indifference,* we may assume that for all $i \in N$, $\sigma_i = i$. First, θ can be obtained by misrepresenting the waiting cost of the last agent downward from θ^2. By *strategyproofness for extreme cost agents,* $t_3 = \theta_2$. Alternatively, θ can be obtained by misrepresenting the waiting cost of the first agent upward from θ^3. Once again, by *strategyproofness for extreme cost agents,* $t_1 = -\theta_2$. Finally, by *budget balance,* $t_2 = 0$. Altogether, we have $t = (-\theta_2, 0, \theta_2) = t^B$, implying that on the class of three-agent problems, $\varphi = \varphi^B$. By *3-subgroup additivity,* we conclude that $\varphi = \varphi^B$ for all problems.

(ii) It is obvious that the symmetrically balanced VCG rule φ^B satisfies *queue-efficiency, budget balance, Pareto indifference, 3-equal treatment of equals, 3-positional equivalence,* and *3-subgroup additivity*. To prove the converse statement, let φ a rule satisfying the six axioms. We begin with a three-agent problem. To simplify the notation, we assume without loss of generality that $N \in \mathcal{N}$ and $\theta \in \mathcal{Q}^N$ be such that $N = \{1, 2, 3\}$ and $\theta_1 \geq \theta_2 \geq \theta_3$. Let $(\sigma, t) \in \varphi(\theta)$. By *queue-efficiency* and *Pareto indifference,* we may assume that for all $i \in N$, $\sigma_i = i$. By *3-positional equivalence,* one of the three agents should be chosen to satisfy the indifference requirements. Together with *budget balance,* we obtain the following three possibilities:

(1) $\sigma_0 = 1$, $t_1 = -\theta_1$, $t_2 = 0$, and $t_3 = \theta_1$.
(2) $\sigma_0 = 2$, $t_1 = -\theta_2$, $t_2 = 0$, and $t_3 = \theta_2$. (Therefore, $t = t^B$.)
(3) $\sigma_0 = 3$, $t_1 = -\theta_3$, $t_2 = 0$, and $t_3 = \theta_3$.

First, consider $\theta' \in \mathcal{Q}^N$ such that $\theta'_1 > \theta'_2 = \theta'_3$. Let $(\sigma', t') \in \varphi(\theta')$. By *queue-efficiency* and *Pareto indifference,* we may assume that for all $i \in N$, $\sigma'_i = i$. If the transfer is chosen to satisfy condition (1), then

$$u_2(\sigma'_2, t'_2; \theta'_2) = -\theta'_2 < -\theta'_3 + (\theta'_1 - \theta'_3) = -2\theta'_3 + \theta'_1 = u_3(\sigma'_3, t'_3; \theta'_3),$$

which violates *3-equal treatment of equals* between agents 2 and 3. Therefore, condition (1) is ruled out. Similarly, consider $\theta'' \in \mathcal{Q}^N$ such that $\theta''_1 = \theta''_2 > \theta''_3$. Let $(\sigma'', t'') \in \varphi(\theta'')$. By *queue-efficiency* and *Pareto indifference,* we may assume that for all $i \in N$, $\sigma''_i = i$. If the transfer is chosen to satisfy condition (3), then

$$u_1(\sigma''_1, t''_1; \theta''_1) = -\theta''_3 > -\theta''_2 = u_2(\sigma''_2, t''_2; \theta''_2),$$

which violates *3-equal treatment of equals* between agents 1 and 2. Therefore, condition (3) is also ruled out. On the other hand, it is easy to check that condition (2) is compatible with *3-equal treatment of equals*. Therefore, on the class of three-agent problems, if a rule φ satisfies *queue-efficiency, budget balance, Pareto indifference, 3-equal treatment of equals,* and *3-positional equivalence,* then $\varphi = \varphi^B$. By *3-subgroup additivity,* we conclude that $\varphi = \varphi^B$ for all problems. □

Comparing the two characterizations of the symmetrically balanced VCG rule in Theorem 8.3, it is obvious that given *queue-efficiency, budget balance, Pareto indifference, 3-equal treatment of equals,* and *3-subgroup additivity, strategyproofness for extreme cost agents* and *3-positional equivalence* are substitutes. Also, it is easy to show that in our Theorem 8.2(i), we can impose *2-positional equivalence* instead of *2-egalitarian equivalence with fixed queue position*.

8.7 Concluding Remarks

In this chapter, we investigate the implications of *subgroup additivity* together with *queue-efficiency, equal treatment of equals,* and *Pareto indifference* to characterize five important rules widely discussed in the queueing literature. These five rules are characterized in the context of queueing problems from different strategic and/or normative considerations. This chapter shows that these rules satisfy a common axiom of *subgroup additivity*. Given *subgroup additivity* of these rules, the characterizations become simpler since what matters now is the set of axioms only on two- or three-agent problems. What our analysis shows is that under *subgroup additivity,* strategic axioms (like weaker versions of *strategyproofness*) and equity axioms (like stronger versions of *2-egalitarian equivalence*) can act as substitutes. It must be noted that if we try to identify rules that satisfy *queue-efficiency, strategyproofness,* and *egalitarian equivalence,* then we do obtain a nonempty family of rules (see Theorem 7.1). However, none of the five rules characterized in this chapter is included in this nonempty family of rules. Therefore, for substitutability between strategic and equity axioms, *subgroup additivity* plays a significant role.

References

Chun, Y. (2006). A pessimistic approach to the queueing problem. *Mathematical Social Sciences, 51*, 171–181.

Chun, Y., & Mitra, M. (2014). Subgroup additivity in the queueing problem. *European Journal of Operational Research, 238*(1), 281–289.

Chun, Y., Mitra, M., & Mutuswami, S. (in press). A characterization of the symmetrically balanced VCG rule in the queueing problem. Games and Economic Behavior.

Cres, M., & Moulin, H. (2001). Scheduling with opting out: Improving upon random priority. *Operations Research, 49*, 565–577.

Foster, J., Greer, J., & Thorbecke, E. (1984). A class of decomposable poverty measures. *Econometrica, 52*, 761–766.

Kayi, C., & Ramaekers, E. (2010). Characterizations of Pareto-efficient, fair, and strategy-proof allocation rules in queueing problems. *Games and Economic Behavior, 68*, 220–232.

Kayi, C., & Ramaekers, E. (in press). Corrigendum to Characterizations of Pareto-efficient, fair, and strategy-proof allocation rules in queueing problems [Games and Economic Behavior 68:220–232]. Games and Economic Behavior.

Maniquet, F. (2003). A characterization of the Shapley value in queueing problems. *Journal of Economic Theory, 109*, 90–103.

Mitra, M. (2001). Mechanism design in queueing problems. *Economic Theory, 17*, 277–305.

Mitra, M., & Mutuswami, S. (2011). Group strategyproofness in queueing models. *Games and Economic Behavior, 72*, 242–254.

Moulin, H. (2007). Minimizing the worst slowdown: Off-line and on-line. *Operations Research, 55*, 876–889.

Pazner, E., & Schmeidler, D. (1978). Egalitarian equivalent allocations: A new concept of equity. *Quarterly Journal of Economics, 92*, 671–687.

Peleg, B. (1986). On the reduced game property and its converse. *International Journal of Game Theory, 15*, 187–200.

Shapley, L. S. (1953). A Value for n-person Games. In H. W. Kuhn & A. W. Tucker (Eds.), *Contributions to the theory of games II*. Annals of mathematics studies (Vol. 28, pp. 307–317). Princeton: Princeton University Press.

Suijs, J. (1996). On incentive compatibility and budget balancedness in public decision making. *Economic Design, 2*, 193–209.

Thomson, W. (2004). Consistency and its converse: An elementary introduction. University of Rochester, Rochester Center for Economic Research Working Paper.

Chapter 9
A Noncooperative Approach

9.1 Introduction

The queueing problem can be solved by adopting a strategic approach which builds up a natural and intuitive bargaining protocol such that players can negotiate among themselves to resolve the queueing conflicts. Exploring the bargaining approach for queueing problems is not only important in its own right as providing a new toolbox and contributing to an open area of the problem, but has more significant implications: First, it helps understand the strategic features of the rules and makes a fresh review of their plausibility. Next, we can make a better comparison between different rules and associate axiomatic properties with individuals' rational behavior. Furthermore, new insights on fundamental and methodological issues can be developed.

Two well-known rules for the queueing problem are introduced by applying solutions developed for TU (transferable utility) games (see Chap. 3 for details). The minimal transfer rule (Maniquet 2003) corresponds to the Shapley value (Shapley 1953) of TU games when the worth of a coalition is defined to be the minimum waiting cost incurred by its members under the optimistic assumption that they are served before non-coalitional members. On the other hand, the maximal transfer rule (Chun 2006) corresponds to the Shapley value when the worth of a coalition is defined to be the minimum waiting cost incurred by its members under the pessimistic assumption that they are served after non-coalitional members. Given the connection between the Shapley value for TU games and the minimal and the maximal transfer rules for queueing problems, various bargaining protocols implementing the Shapley value in the literature (Gul 1989; Hart and Mas-Colell 1996; Ju 2013; Ju and Wettstein 2009; Pérez-Castrillo and Wettstein 2001) offer a venue enabling us to construct noncooperative mechanisms to implement rules for queueing problems.

However, this task is not straightforward, especially when considering that the potential mechanism needs to match the underlying context of the queueing

© Springer International Publishing Switzerland 2016

Y. Chun, *Fair Queueing*, Studies in Choice and Welfare,
DOI 10.1007/978-3-319-33771-5_9

problem. Unlike in a TU game where a player's stand-alone value is fixed, a player's stand-alone value is not well defined in the queueing context. It is conventional in the implementation literature on TU games that when the proposal of a player is rejected in multilateral bargaining, she is left with her stand-alone value, which is not affected by the other players' coalitional behavior and does not affect the other players' payoffs. However, for queueing games, a player's utility is necessarily affected by how the others queue, and where to position this player also affects the utility of other players. Take, for example, the queueing game of Maniquet (2003) which is defined in the optimistic perspective. In this game, a player's stand-alone value is defined by having this player be served first before anyone else. However, it is impossible to make every player be served first simultaneously in order to apply this stand-alone value in any bargaining protocol that could be associated with queuing problems. Similar arguments carry over to the pessimistic queueing games introduced by Chun (2006). Moreover, if we directly follow the protocol of Pérez-Castrillo and Wettstein (2001) implementing the Shapley value, it would fail to implement the minimal transfer rule since the underlying queueing game violates zero monotonicity.[1] Hence, the bargaining protocol for the Shapley value cannot be directly applied to the queueing context.

The other challenge lies in the incentive design for players to form an efficient queue while accepting to make transfers as directed by the two transfer rules. Given the queueing context, it should be endogenously designed into the mechanism such that players find themselves being better off by building up an efficient queue, rather than imposing conditions like super-additivity[2] or zero monotonicity.

In this chapter, we introduce two noncooperative mechanisms that naturally fit into the context of queueing problems and retain the main feature for TU games, but well overcome the challenges mentioned above. Players can resolve queueing conflicts by themselves in a decentralized way and guarantee an efficient queue to be formed in equilibrium. These mechanisms have a unique subgame perfect equilibrium (SPE) outcome, which coincides with the payoff vector assigned by one of the two well-known rules for the queueing problem, either the maximal transfer rule or the minimal transfer rule. By keeping the basic construct and adjusting certain details of the bargaining protocols, we can construct other mechanisms that offer strategic foundations for alternative rules of queueing problems. This provides a common platform to study and compare solution concepts for queueing problems and may help investigate new rules.

This chapter is organized as follows.[3] In Sects. 9.2 and 9.3, we construct two noncooperative mechanisms and show that each mechanism has a unique subgame

[1]A game v satisfies *zero monotonicity* if there are no negative externalities when a single player joins a coalition. That is, for all $S \subset N$ and all $i \notin S$, $v(S \cup \{i\}) \geq v(S) + v(\{i\})$.

[2]A game v satisfies *super-additivity* if there are no negative externalities when two disjoint coalitions are merged together. That is, for all $S, T \subset N$ such that $S \cap T = \emptyset$, $v(S \cup T) \geq v(S) + v(T)$.

[3]This chapter is based mainly on Ju et al. (2014a,b). Excerpts from Ju et al. (2014b) are reprinted with kind permission of Elsevier.

perfect equilibrium outcome which is the allocation assigned by the maximal transfer rule and the minimal transfer rule, respectively. In Sect. 9.4, we provide a mechanism to implement the average of the maximal and the minimal transfer rules. Finally, in Sect. 9.5, we offer a robustness study of the results, which shows that to a great extent the ordering of rejected players does not affect the equilibrium outcome.

9.2 The First-Served Mechanism and the Maximal Transfer Rule

In Sects. 9.2 and 9.3, we implement the maximal and the minimal transfer rules by constructing two bargaining games that well fit into the context of queueing problems. We thus hope to provide a strategic perspective to evaluate and compare the two rules. Here players are assumed to be risk neutral and expected utility maximizers.

The first game, called the *first-served mechanism*, which implements the maximal transfer rule, can be described informally as follows. At stage F-1, all players participate in a multi-bidding auction to compete for the first position of a queue. In this auction, each player bids by submitting an $(n-1)$-tuple of numbers (positive or negative), one number for each player (excluding herself). A positive number means a payment she makes to another player and a negative number means a compensation she asks for from another player. The player whose net bid (the difference between the sum of bids made by the player and the sum of bids the other players made to her) is the highest wins the first position while making the payment or receiving the compensation, as per the corresponding bid she makes. At stage F-2, the winner has two options. She can either take up the first position by herself or sell it to other players. If she decides to take up the position by herself, then the rest of the players play the game again from the first stage to bargain over the positions after her. If she decides to sell the position, then this sale cannot be a bilateral one because where to locate the winner after the sale affects other players' positions. Therefore, selling the first position is naturally an all-party negotiation process. That is, the winner makes an overall proposal that consists of a queue assigning positions to all players and a vector of transfers specifying the amount each player is supposed to pay or receive. Stage F-3 is to approve or disapprove the proposal. The proposal is accepted if all the other players agree. In case of acceptance, the proposal is implemented so that the queue is formed with transfers in effect to all players. In case of rejection, the proposer loses any say about the queue, but she retains the first position,[4] although she incurs no additional transfers, i.e., neither pay to others nor receive any compensation from others (except for the bids made at

[4]Indeed this option makes the choice of taking up the first position at stage F-2 strategically redundant. Yet it seems natural and logical for the winner to have the right to take up the position without proceeding to the next stage.

stage F-1). Meanwhile, all players except for the rejected proposer start new round of negotiation using the same rule.

Now we formally describe the first-served mechanism.

First-Served Mechanism If there is only one player, $N = \{i\}$, she simply is served first (also last), and the default queue is trivially efficient while no transfer is made. So this player's utility is $u_i(\sigma, t) = -(\sigma_i - 1)\theta_i + t_i = -0\theta_i + 0 = 0$, which is independent of θ_i. When the player set $N = \{1, \ldots, n\}$ consists of two or more players, the mechanism is defined for any set of (active) players $S \subseteq N$, recursively starting with $S = N$.

Stage F-1. Bidding for the first position: Each player $i \in S$ makes $s-1$ (where $s = |S|$ is the cardinality of coalition S) bids $b_j^i \in \mathbf{R}$, one to every player $j \in S\backslash\{i\}$. For each player $i \in S$, define the *net bid* of player i by $B^i = \sum_{j\in S\backslash\{i\}} b_j^i - \sum_{j\in S\backslash\{i\}} b_i^j$. Let player $i_s = \text{argmax}_{i\in S} B^i$. In case of a nonunique maximizer, we choose with equal probability any of these maximal bidders to be the "winner." Once the winner i_s is chosen, she pays every player $j \in S\backslash\{i_s\}$ her bid $b_j^{i_s}$.

Stage F-2. Taking up or selling the position: The winner i_s decides to either take up the first position by herself or sell it to the others. If winner i_s takes up the position by herself, then she is located before all players of $S\backslash\{i_s\}$ but after $N\backslash S$, i.e., at position $|N| - |S| + 1$. Her final utility is $-(|N| - |S|)\theta_i - \sum_{j\in S\backslash\{i_s\}} b_j^{i_s} + \sum_{r=s+1}^{n} b_{i_s}^{i_r}$, where $(|N| - |S|)\theta_i$ is her waiting cost at this position, $b_j^{i_s}$ is the bid she pays to player $j \in S \setminus \{i_s\}$, and $b_{i_s}^{i_r}$, $r \in \{s+1, \ldots, n\}$, are the bids she received from the previously rejected proposers i_r, $r \in \{s+1, \ldots, n\}$. Moreover, stage F-3 is not evoked, but all players in S other than player i_s proceed again from stage F-1 where the set of active players is $S\backslash\{i_s\}$. If the winner i_s decides to sell the position, then she makes a proposal $(\sigma(S), (t_j)_{j\in S})$ consisting of a queue $\sigma(S) \in \Sigma(S)$ and a vector of transfers $(t_j)_{j\in S} \in \mathbf{R}^S$ such that $\sum_{j\in S} t_j \leq 0$. (This offer is additional to the bids paid at stage F-1.) The game continues to stage F-3.

Stage F-3. Approving or disapproving a proposal: The players in S other than player i_s, sequentially, either accept or reject the proposal. If at least one player rejects the proposal, then it is rejected. Otherwise, it is accepted.

(i) If the proposal is rejected, all players in S other than player i_s go back to stage F-1 where the set of active players is $S\backslash\{i_s\}$. Meanwhile, player i_s simply falls back to her default position: the position in front of the queue of $S\backslash\{i_s\}$ but after $N\backslash S$. Consequently, player i_s receives her final utility $-(|N| - |S|)\theta_i - \sum_{j\in S\backslash\{i_s\}} b_j^{i_s} + \sum_{r=s+1}^{n} b_{i_s}^{i_r}$. Note that she would receive the same utility if she decided in stage F-2 to take up her position.

(ii) If the proposal is accepted, then we have to distinguish between two cases, $S = N$ and $S \neq N$. If $S = N$, all players agree with the proposer i_n on her proposal $(\sigma(N), (t_j)_{j\in N})$, and the game ENDS. Each player $j \in N\backslash\{i_n\}$ receives $-(\sigma_j(N) - 1)\theta_j + b_j^{i_n} + t_j$, and player i_n receives $-(\sigma_{i_n}(N) - 1)\theta_{i_n} - \sum_{j\in N\backslash\{i_n\}} b_j^{i_n} + t_{i_n}$, where $t_{i_n} = -\sum_{j\in N\backslash\{i_n\}} t_j$. If $S \neq N$, all players in S agree with the proposer i_s on her proposal $(\sigma(S), (t_j)_{j\in S})$, and the game ENDS.

Each player $j \in S\backslash\{i_s\}$ receives $-(|N| - |S| + \sigma_j(S) - 1)\theta_j + t_j$, and with the bids made by player i_s and all other previously rejected proposers, player j's final utility is $-(|N| - |S| + \sigma_j(S) - 1)\theta_j + \sum_{k=s}^{n} b_j^{i_k} + t_j$. Player i_s receives $-(|N|-|S|+\sigma_{i_s}(S)-1)\theta_{i_s} - \sum_{j \in S\backslash\{i_s\}} b_j^{i_s} + t_{i_s}$, where $t_{i_s} = -\sum_{j \in S\backslash\{i_s\}} t_j$, and adding the bids player i_s received from the previously rejected proposers, her final utility is $-(|N| - |S| + \sigma_{i_s}(S) - 1)\theta_{i_s} - \sum_{j \in S\backslash\{i_s\}} b_j^{i_s} + \sum_{k=s+1}^{n} b_{i_s}^{i_k} + t_{i_s}$.

We show that for any $N \in \mathscr{N}$ and any $\theta \in \mathscr{Q}^N$, the first-served mechanism has a unique subgame perfect equilibrium (SPE) outcome, which coincides with the payoff vector assigned by the maximal transfer rule. For all $N \in \mathscr{N}$, all $i \in N$, and all $S \subseteq N\backslash\{i\}$, consider an efficient queue σ^* for $S \cup \{i\}$, i.e., $\sigma^* \in \mathit{Eff}(\theta_{S\cup\{i\}})$, and an efficient queue σ^{**} for S, i.e., $\sigma^{**} \in \mathit{Eff}(\theta_S)$. Then, the marginal contribution of player $i \in N$ to coalition $S \subseteq N\backslash\{i\}$ in the pessimistic queueing game v_P equals

$$
\begin{aligned}
& v_P(S \cup \{i\}) - v_P(S) \\
&= -\sum_{k \in S\cup\{i\}} (|N| - |S| - 1 + \sigma_k^* - 1)\theta_k - \left(-\sum_{j \in S} (|N| - |S| + \sigma_j^{**} - 1)\theta_j \right) \\
&= \sum_{j \in S} (|N| - |S| + \sigma_j^{**})\theta_j - \sum_{k \in S\cup\{i\}} (|N| - |S| - 1 + \sigma_k^*)\theta_k + \theta_i \\
&= \left(\sum_{j \in P_i(\sigma^*)} (|N| - |S| + \sigma_j^*)\theta_j + \sum_{j \in F_i(\sigma^*)} (|N| - |S| + \sigma_j^* - 1)\theta_j \right) \qquad (9.1) \\
&\quad - \left(\sum_{j \in P_i(\sigma^*)} (|N| - |S| - 1 + \sigma_j^*)\theta_j + (|N| - |S| - 1 + \sigma_i^*)\theta_i \right. \\
&\quad \left. + \sum_{j \in F_i(\sigma^*)} (|N| - |S| - 1 + \sigma_j^*)\theta_j \right) + \theta_i \\
&= (2 - |N| + |S| - \sigma_i^*)\theta_i + \sum_{j \in P_i(\sigma^*)} \theta_j,
\end{aligned}
$$

where the first equality follows by definition of v_P, and the third equality follows from the fact that $\sigma_j^{**} = \sigma_j^* + 1$ for all $j \in P_i(\sigma^*)$ and $\sigma_j^{**} = \sigma_j^*$ for all $j \in F_i(\sigma^*)$. That is, with the joining of player i to coalition S, construction of the efficient queue requires those who have higher waiting costs than θ_i to be served before player i and those who have lower waiting costs than θ_i to be served after player i. Hence, compared to the total cost within S, player i's marginal contribution (in terms of cost) is

$$
-\{(|N| - (|S| + 1)) + (\sigma_i^* - 1)\}\theta_i + \sum_{j \in P_i(\sigma^*)} \theta_j = (2 - |N| + |S| - \sigma_i^*)\theta_i + \sum_{j \in P_i(\sigma^*)} \theta_j.
$$

We begin with the following lemma.

Lemma 9.1 *For all $N \in \mathcal{N}$, all $\theta \in \mathcal{Q}^N$, all $i \in N$, and all $S \subseteq N\backslash\{i\}$,*

$$v_P(S \cup \{i\}) - v_P(S) \geq -(|N| - |S| - 1)\theta_i.$$

Proof Let $N \in \mathcal{N}$, $\theta \in \mathcal{Q}^N$, $i \in N$, and $S \subseteq N\backslash\{i\}$. Also, let $T \subseteq S$ be the set of players whose unit waiting costs are greater than θ_i and $S\backslash T$ be the set of players whose unit waiting costs are less than or equal to θ_i. That is, $\theta_j > \theta_i \geq \theta_k$ for all $j \in T$ and all $k \in S\backslash T$. In an efficient queue σ^* for $S \cup \{i\}$, player i will be served after players in T but before anyone else in the queue σ^*, so that $\sigma_i^* = |T| + 1$.[5] Thus,

$$\begin{aligned}
v_P(S \cup \{i\}) - v_P(S) &= (2 - |N| + |S| - \sigma_i^*)\theta_i + \sum_{j \in P_i(\sigma^*)} \theta_j \\
&= (2 - |N| + |S| - (|T| + 1))\,\theta_i + \sum_{j \in T} \theta_j \\
&= -(|N| - |S| + |T| - 1)\theta_i + \sum_{j \in T} \theta_j \\
&\geq -(|N| - |S| + |T| - 1)\theta_i + |T|\theta_i \\
&= -(|N| - |S| - 1)\theta_i.
\end{aligned}$$

□

The inequality in Lemma 9.1 is strict if $\theta_i < \theta_j$ for at least one $j \in S \setminus \{i\}$. The implication of this lemma is rather intuitive: it pays for player i to join coalition S and form an efficient queue rather than taking up a position in front of coalition S since the cost for the coalition $S \cup \{i\}$ is not more than the cost of coalition S plus the cost of player i when she is served in the position before coalition S. The right-hand side of the inequality of the lemma is the utility of player i when she takes up the position $\sigma_i = |N| - |S|$. Although this result shares a similar feature as the zero-monotonicity property for TU games, there is an important difference which lies in the fact that the zero-monotonicity property makes a comparison between the marginal contribution of a player and her stand-alone worth. Note that the stand-alone worth of player i in this game is $-(|N| - 1)\theta_i$. Moreover, the cost $-(|N| - |S| - 1)\theta_i$ depends on the size of coalition S, while the stand-alone worth $v_P(\{i\}) = -(|N| - 1)\theta_i$ does not.

[5] Note that the position of player i may not be $|T| + 1$ if there is a player $j \in S$ with $\theta_j = \theta_i$. Since the choice of an efficient queue has no effect on $v_P(S \cup \{i\}) - v_P(S)$, we can take an efficient queue σ^* with $\sigma_i^* = |T| + 1$.

Theorem 9.1 *For all $N \in \mathcal{N}$ and all $\theta \in \mathcal{Q}^N$, the first-served mechanism has a unique subgame perfect equilibrium (SPE) outcome, which coincides with the payoff vector assigned by the maximal transfer rule $\varphi^C(\theta)$.*

Proof Let $N \in \mathcal{N}$ be such that $N = \{1, \ldots, n\}$ and $\theta \in \mathcal{Q}^N$. The proof proceeds by induction on the number of players n. The induction assumption is that whenever the mechanism is used by n players with a given vector of unit waiting costs, it implements the maximal transfer rule of this queueing problem. It is easy to see that the theorem holds for $n = 1$. We assume that it holds for all $m \leq n - 1$ and show that it is satisfied for n.

First we show that the allocation $\varphi^C(\theta)$ assigned by the maximal transfer rule is an SPE outcome. We explicitly construct an SPE that yields $\varphi^C(\theta)$ as the SPE outcome. Consider the following strategies, which the players would follow in any (sub)game they participate in (we describe it for the whole set of players, N, but similar strategies are followed by any player in $S \subseteq N$ that is called upon to play the game, with S replacing N):

At stage F-1, each player $i \in N$ announces

$$b_j^i = \left(-(\sigma_j^* - 1)\theta_j + \sum_{k \in P_j(\sigma^*)} \frac{\theta_k}{2} - (|N| - \sigma_j^*)\frac{\theta_j}{2}\right)$$
$$- \left(-(1 + \sigma_j^{**} - 1)\theta_j + \sum_{k \in P_j(\sigma^{**})} \frac{\theta_k}{2} - (|N| - 1 - \sigma_j^{**})\frac{\theta_j}{2}\right)$$

where $\sigma^* \in \mathit{Eff}(\theta)$ and $\sigma^{**} \in \mathit{Eff}(\theta_{N\setminus\{i_n\}})$ for all $j \in N\setminus\{i\}$.

At stage F-2, the proposer i_n ("winner" of the bidding in stage F-1) adopts the option of selling the position instead of taking it up by herself and makes a proposal (σ^*, t) such that $\sigma^* \in \mathit{Eff}(\theta)$ and for all $j \in N\setminus\{i_n\}$,

$$t_j = (\sigma_j^* - 1)\theta_j + \left(-(1 + \sigma_j^{**} - 1)\theta_j + \sum_{k \in P_j(\sigma^{**})} \frac{\theta_k}{2} - (|N| - 1 - \sigma_j^{**})\frac{\theta_j}{2}\right)$$

where $\sigma^{**} \in \mathit{Eff}(\theta_{N\setminus\{i_n\}})$.

At stage F-3, any player $j \in N\setminus\{i_n\}$ accepts any proposal (σ, t) if $\sigma \in \mathit{Eff}(N)$ and

$$t_j \geq (\sigma_j - 1)\theta_j + \left(-(1 + \sigma_j^{**} - 1)\theta_j + \sum_{k \in P_j(\sigma^{**})} \frac{\theta_k}{2} - (|N| - 1 - \sigma_j^{**})\frac{\theta_j}{2}\right)$$

where $\sigma^{**} \in \mathit{Eff}(\theta_{N\setminus\{i_n\}})$, and rejects it otherwise. (Note that the right-hand side is the same for any $\sigma^{**} \in \mathit{Eff}(\theta_{N\setminus\{i_n\}})$.)

To verify that the above strategies indeed constitute an SPE and yield the allocation assigned by the maximal transfer rule, we first look at stage F-3. Suppose the proposal of player i_n is rejected. From the induction hypothesis, we know that for the remaining players in $N\backslash\{i_n\}$, a corresponding efficient queue is formed in the unique SPE outcome of that subgame and the resulting utility to every player $j \in N\backslash\{i_n\}$ is $-(1+\sigma_j^{**}-1)\theta_j + \sum_{k\in P_j(\sigma^{**})} \frac{\theta_k}{2} - (|N|-1-\sigma_j^{**})\frac{\theta_j}{2}$, where $\sigma^{**} \in Eff(\theta_{N\backslash\{i_n\}})$, which is the reservation utility for player j when she considers any proposal made by player i_n. If player j accepts the proposal made by player i_n, then she is located at position σ_j with transfer t_j so that she receives the utility $-(\sigma_j - 1)\theta_j + t_j$. Apparently, only when this utility is no less than her reservation utility, she accepts the proposal, which gives rise to $t_j \geq (\sigma_j - 1)\theta_j + \left(-(1+\sigma_j^{**}-1)\theta_j + \sum_{k\in P_j(\sigma^{**})} \frac{\theta_k}{2} - (|N|-1-\sigma_j^{**})\frac{\theta_j}{2}\right)$. Note that with t_j as specified above, at this stage, player j's utility is guaranteed to be no less than her reservation utility regardless of the queue proposed by player i_n.

Now consider stage F-2. Obviously, for any proposed σ, player i_n does not make any player $j \in N\backslash\{i_n\}$ an offer t_j that is strictly higher than $(\sigma_j - 1)\theta_j + \left(-(1+\sigma_j^{**}-1)\theta_j + \sum_{k\in P_j(\sigma^{**})} \frac{\theta_k}{2} - (|N|-1-\sigma_j^{**})\frac{\theta_j}{2}\right)$. In the meantime, player i_n would not lower the offer t_j for any player $j \in N\backslash\{i_n\}$' to be strictly less than $(\sigma_j - 1)\theta_j + \left(-(1+\sigma_j^{**}-1)\theta_j + \sum_{k\in P_j(\sigma^{**})} \frac{\theta_k}{2} - (|N|-1-\sigma_j^{**})\frac{\theta_j}{2}\right)$ since it would lead her proposal to be rejected and then to be served at first. Being served first implies 0 waiting cost for player i_n at this stage, but there is an incentive for player i_n to make an acceptable proposal as presented at stage F-2 since her payoff from an accepted proposal is

$$\begin{aligned}
&-(\sigma_{i_n}^* - 1)\theta_{i_n} - \sum_{j\in N\backslash\{i_n\}} t_j \\
&= -(\sigma_{i_n}^* - 1)\theta_{i_n} - \sum_{j\in N\backslash\{i_n\}} (\sigma_j^* - 1)\theta_j \\
&\quad - \sum_{j\in N\backslash\{i_n\}} \left\{-(1+\sigma_j^{**}-1)\theta_j + \sum_{k\in P_j(\sigma^{**})} \frac{\theta_k}{2} - (|N|-1-\sigma_j^{**})\frac{\theta_j}{2}\right\} \\
&= -\sum_{i\in N}(\sigma_i^* - 1)\theta_i - \sum_{j\in N\backslash\{i_n\}} \{-(1+\sigma_j^{**}-1)\theta_j\} \\
&\quad - \sum_{j\in N\backslash\{i_n\}} \left\{\sum_{k\in P_j(\sigma^{**})} \frac{\theta_k}{2} - (|N|-1-\sigma_j^{**})\frac{\theta_j}{2}\right\} \\
&= v_P(N) - v_P(N\backslash\{i_n\}) \\
&\geq -(|N| - (|N|-1) - 1)\theta_{i_n} \\
&= 0,
\end{aligned}$$

where the third equality follows from the definition of v_P and the fact that by definition of an efficient allocation, $\sum_{j\in N\setminus\{i_n\}}\left(\sum_{k\in P_j(\sigma^{**})}\frac{\theta_k}{2}-(|N|-1-\sigma_j^{**})\frac{\theta_j}{2}\right)=0$, and the inequality follows from Lemma 9.1.

To verify that the strategies restricted to stage F-1 constitute an SPE, note that all net bids equal zero, which follows from the fact that $b_j^i=\phi_j(v_P)-\phi_j(v_P|_{N\setminus\{i\}})$ where $v_P|_{N\setminus\{i\}}(S)=v_P(S)$ for all $S\subseteq N\setminus\{i\}$ and by the balanced contributions property of the Shapley value (Myerson 1980; van den Brink and Chun 2012).[6] To show that a change in the bids made by player i cannot increase that player's payoff, we consider the following two cases: First, if player i changes the vector of her bids so that another player becomes the proposer, this does not change her resulting utility, which would still equal that given by the maximal transfer rule (i.e., her allocation assigned by the Shapley value of the pessimistic queueing game v_P). Second, if she changes the vector of her bids and following it she is still the proposer with positive probability, it must be that her total bid ($\sum_{j\in N\setminus\{i\}}b_j^i$) does not decline, which again means her payoff cannot improve. That is, any deviation of the bidding strategy of player i specified at stage F-1 cannot improve her payoff. Hence, no player has an incentive to change her bid, showing that the given strategy profile is an SPE.

The proof that any SPE yields the allocation specified by the maximal transfer rule proceeds by a series of claims.

Claim F-1 At stage F-3, in any SPE, any player $j\in N\setminus\{i_n\}$ accepts any proposal (σ,t) such that $\sigma\in\Sigma(N)$ if

$$t_j\geq(\sigma_j-1)\theta_j+\left(-(1+\sigma_j^{**}-1)\theta_j+\sum_{k\in P_j(\sigma^{**})}\frac{\theta_k}{2}-(|N|-1-\sigma_j^{**})\frac{\theta_j}{2}\right),$$

and rejects it if

$$t_j<(\sigma_j-1)\theta_j+\left(-(1+\sigma_j^{**}-1)\theta_j+\sum_{k\in P_j(\sigma^{**})}\frac{\theta_k}{2}-(|N|-1-\sigma_j^{**})\frac{\theta_j}{2}\right)$$

where $\sigma^{**}\in \mathit{Eff}(\theta_{N\setminus\{i_n\}})$.

Proof This claim follows directly from the induction assumption. □

Claim F-2

(i) If $v_P(N)-v_P(N\setminus\{i_n\})>-(|N|-(|N|-1)-1)\theta_{i_n}=0$, the only SPE of the game that starts at stage F-2 is the following. At stage F-2, player i_n chooses the option of selling the position instead of taking it up by herself and makes a

[6]A value ϕ satisfies the *balanced contributions property* if $\phi_i(v)-\phi_i(v|_{N\setminus\{j\}})=\phi_j(v)-\phi_j(v|_{N\setminus\{i\}})$ for all $v\in\Gamma^N$ and all $i,j\in N$. Section 4.4 investigates the implication of this property in the context of queueing.

proposal (σ^*, t) such that $\sigma^* \in \mathit{Eff}(\theta)$ and

$$t_j = (\sigma_j^* - 1)\theta_j + \left(-(1 + \sigma_j^{**} - 1)\theta_j + \sum_{k \in P_j(\sigma^{**})} \frac{\theta_k}{2} - (|N| - 1 - \sigma_j^{**})\frac{\theta_j}{2}\right)$$

where $\sigma^{**} \in \mathit{Eff}(\theta_{N\backslash\{i_n\}})$ to each player $j \in N\backslash\{i_n\}$. At stage F-3, each player $j \in N\backslash\{i_n\}$ accepts any proposal (σ, t) such that $\sigma \in \Sigma(N)$ if

$$t_j \geq (\sigma_j - 1)\theta_j + \left(-(1 + \sigma_j^{**} - 1)\theta_j + \sum_{k \in P_j(\sigma^{**})} \frac{\theta_k}{2} - (|N| - 1 - \sigma_j^{**})\frac{\theta_j}{2}\right),$$

where $\sigma^{**} \in \mathit{Eff}(\theta_{N\backslash\{i_n\}})$, and rejects it otherwise.

(ii) If $v_P(N) - v_P(N\backslash\{i_n\}) = 0$, there exist other SPEs in addition to the one described above. In fact, any set of the following strategies also constitutes an SPE: at stage F-2, player i_n either takes up the first position by herself or sells the position by making a proposal (σ, t) such that $\sigma \in \Sigma(N)$ and to some player $j \neq i_n$,

$$t_j < (\sigma_j - 1)\theta_j + \left(-(1 + \sigma_j^{**} - 1)\theta_j + \sum_{k \in P_j(\sigma^{**})} \frac{\theta_k}{2} - (|N| - 1 - \sigma_j^{**})\frac{\theta_j}{2}\right)$$

where $\sigma^{**} \in \mathit{Eff}(\theta_{N\backslash\{i_n\}})$; and at stage F-3, player j rejects any proposal

$$t_j \leq (\sigma_j - 1)\theta_j + \left(-(1 + \sigma_j^{**} - 1)\theta_j + \sum_{k \in P_j(\sigma^{**})} \frac{\theta_k}{2} - (|N| - 1 - \sigma_j^{**})\frac{\theta_j}{2}\right).$$

In any SPE of this subgame, the final payoffs to players i_n and $j \neq i_n$ are $-(\sigma_{i_n}^* - 1)\theta_{i_n} - \sum_{j \in N\backslash\{i_n\}} b_j^{i_n} - \sum_{j \in N\backslash\{i_n\}} t_j$ and $-(\sigma_j^* - 1)\theta_j + b_j^{i_n} + t_j$, respectively, where $t_j = (\sigma_j^* - 1)\theta_j + \left(-(1 + \sigma_j^{**} - 1)\theta_j + \sum_{k \in P_j(\sigma^{**})} \frac{\theta_k}{2} - (|N| - 1 - \sigma_j^{**})\frac{\theta_j}{2}\right)$.

Proof For the case of $v_P(N) - v_P(N\backslash\{i_n\}) > 0$, one can verify the argument by the induction assumption and Lemma 9.1. For the case of $v_P(N) - v_P(N\backslash\{i_n\}) = 0$, one can obviously see that player i_n would be indifferent between taking up the first position and making an acceptable proposal, with the first option being equivalent to making an unacceptable proposal and then having it be rejected, which would all yield the same payoff to player i_n. Note that in this case $\theta_{i_n} > \theta_j$ for all $j \in N\backslash\{i_n\}$ due to Lemma 9.1, and by the induction hypothesis, taking up the first position by player i_n still yields an efficient queue. □

Claim F-3 In any SPE, $B^i = B^j$ for all $i, j \in N$ and hence $B^i = 0$ for all $i \in N$.

Claim F-4 In any SPE, each player's payoff is the same regardless of who is chosen as the winner at stage F-1.

Proof The proofs of Claims F-3 and F-4 are the same as in Pérez-Castrillo and Wettstein (2001). □

Claim F-5 In any SPE, the final payoff of each player coincides with her payoff assigned by the maximal transfer rule.

Proof Note that if player i is the proposer, her final payoff is

$$-(\sigma_i^* - 1)\theta_i - \sum_{j \in N\setminus\{i\}} b_j^i - \sum_{j \in N\setminus\{i\}} t_j,$$

where $t_j = (\sigma_j^* - 1)\theta_j + \left(-(1 + \sigma_j^{**} - 1)\theta_j + \sum_{k \in P_j(\sigma^{**})} \frac{\theta_k}{2} - (|N| - 1 - \sigma_j^{**})\frac{\theta_j}{2}\right)$. And if player $j \neq i$ is the proposer, the final payoff of player i is

$$-(\sigma_i^* - 1)\theta_i + b_i^j + \left((\sigma_i^* - 1)\theta_i + \left(-(1 + \sigma_i^{**} - 1)\theta_i + \sum_{k \in P_i(\sigma^{**})} \frac{\theta_k}{2} - (|N| - 1 - \sigma_i^{**})\frac{\theta_i}{2}\right)\right).$$

Therefore, the sum of payoffs to player i over all possible choices of the proposer is

$$\begin{aligned}
&-(\sigma_i^* - 1)\theta_i - \sum_{j \in N\setminus\{i\}} b_j^i \\
&- \sum_{j \in N\setminus\{i\}} \left((\sigma_j^* - 1)\theta_j + \left(-(1 + \sigma_j^{**} - 1)\theta_j + \sum_{k \in P_j(\sigma^{**})} \frac{\theta_k}{2} - (|N| - 1 - \sigma_j^{**})\frac{\theta_j}{2}\right)\right) \\
&+ \sum_{j \neq i} \left(-(\sigma_i^* - 1)\theta_i + b_i^j\right) \\
&+ \sum_{j \neq i} \left((\sigma_i^* - 1)\theta_i + \left(-(1 + \sigma_i^{**} - 1)\theta_i + \sum_{k \in P_i(\sigma^{**})} \frac{\theta_k}{2} - (|N| - 1 - \sigma_i^{**})\frac{\theta_i}{2}\right)\right) \\
= \; &-n(\sigma_i^* - 1)\theta_i \\
&- \sum_{j \in N\setminus\{i\}} \left((\sigma_j^* - 1)\theta_j + \left(-(1 + \sigma_j^{**} - 1)\theta_j + \sum_{k \in P_j(\sigma^{**})} \frac{\theta_k}{2} - (|N| - 1 - \sigma_j^{**})\frac{\theta_j}{2}\right)\right) \\
&+ (n-1)(\sigma_i^* - 1)\theta_i + \sum_{j \neq i} \left(-(1 + \sigma_i^{**} - 1)\theta_i + \sum_{k \in P_i(\sigma^{**})} \frac{\theta_k}{2} - (|N| - 1 - \sigma_i^{**})\frac{\theta_i}{2}\right)
\end{aligned}$$

$$
\begin{aligned}
&= -(\sigma_i^* - 1)\theta_i - \sum_{j \in N\setminus\{i\}} (\sigma_j^* - 1)\theta_j - \sum_{j \in N\setminus\{i\}} (-\sigma_j^{**}\theta_j) \\
&\quad + \sum_{j \in N\setminus\{i\}} \left(-\sigma_i^{**}\theta_i + \sum_{k \in P_i(\sigma^{**})} \frac{\theta_k}{2} - (|N| - 1 - \sigma_i^{**})\frac{\theta_i}{2} \right) \\
&= -(\sigma_i^* - 1)\theta_i + \sum_{k \in P_i(\sigma^*)} \theta_k + \sum_{j \in N\setminus\{i\}} \left(-(|N| - 1)\frac{\theta_i}{2} - \sigma_i^{**}\frac{\theta_i}{2} + \sum_{k \in P_i(\sigma^{**})} \frac{\theta_k}{2} \right) \\
&= -(\sigma_i^* - 1)\theta_i + \sum_{k \in P_i(\sigma^*)} \theta_k - \sum_{j \in N\setminus\{i\}} \frac{\sigma_i^{**}\theta_i}{2} - \frac{(n-1)^2}{2}\theta_i + \sum_{j \in N\setminus\{i\}} \sum_{k \in P_i(\sigma^{**})} \frac{\theta_k}{2} \\
&= -(\sigma_i^* - 1)\theta_i + \sum_{k \in P_i(\sigma^*)} \theta_k - \frac{1}{2}\left(|P_i(\sigma^*)|(\sigma^* - 1)\theta_i + |F_i(\sigma^*)|\sigma^*\theta_i\right) - \frac{(n-1)^2}{2}\theta_i \\
&\quad + |F_i(\sigma^*)| \sum_{k \in P_i(\sigma^*)} \frac{\theta_k}{2} + (|P_i(\sigma^*)| - 1) \sum_{k \in P_i(\sigma^*)} \frac{\theta_k}{2} \\
&= -(\sigma_i^* - 1)\theta_i - \frac{1}{2}\left((n-1)\sigma^*\theta_i - |P_i(\sigma^*)|\theta_i\right) - \frac{(n-1)^2}{2}\theta_i + \sum_{k \in P_i(\sigma^*)} \frac{n\theta_k}{2} \\
&= -(\sigma_i^* - 1)\theta_i - \frac{1}{2}\left((n-1)\sigma^*\theta_i - (\sigma_i^* - 1)\theta_i\right) - \frac{(n-1)^2}{2}\theta_i + \sum_{k \in P_i(\sigma^*)} \frac{n\theta_k}{2} \\
&= -\frac{n}{2}(\sigma_i^* - 2 + n)\theta_i + \sum_{k \in P_i(\sigma^*)} \frac{n\theta_k}{2} \\
&= n\left(-(\sigma_i^* - 1)\theta_i - (n - \sigma_i^*)\frac{\theta_i}{2} + \sum_{k \in P_i(\sigma^*)} \frac{\theta_k}{2} \right) \\
&= n\varphi_i^C(\theta).
\end{aligned}
$$

Together with Claim F-4, we conclude that in any SPE, for any player $i \in N$, her final payoff is $\varphi_i^C(\theta)$. □

9.3 The Last-Served Mechanism and the Minimal Transfer Rule

Next we introduce our second game, called the *last-served mechanism*, which implements the minimal transfer rule. Differently from the first-served mechanism, players compete for the right of being served last in the queue. Alternatively, one can think that players are now demanding compensations for them to be served last, which is in the same light as the ALDB (auctioning the leadership with differentiated bids) mechanism (Moulin 1981). The one with the highest net bid (or lowest net compensation if the bids are negative) is selected as the winner. The winner can decide to take up the last position by herself or sell it to the others. For the latter option, she makes a proposal of a queue and a vector of transfers. If the proposal is rejected, she remains at the last position to be served after all the participating players.

Now we provide a formal description of the last-served mechanism, again based on any set of (active) players $S \subseteq N$.

Last-Served Mechanism If there is only one player, $N = \{i\}$, she simply is served last (also first), and the default queue is trivially efficient while no transfer is made. So this player's utility is $u_i(\sigma, t) = -(\sigma_i - 1)\theta_i + t_i = -0\theta_i + 0 = 0$, which is independent of θ_i. When the player set $N = \{1, \ldots, n\}$ consists of two or more players, the mechanism is defined for any set of (active) players $S \subseteq N$, recursively starting with $S = N$.

Stage L-1. Bidding for the last position: It is the same as stage F-1 of the first-served mechanism, except that the right at the winner's disposal in this mechanism is to be served after, instead of before, the rest of the active players.

Stage L-2. Taking up or selling the position: The winner i_s decides to either take up the last position by herself or sell it to the others. If winner i_s takes up the position by herself, she is indeed located at the sth position, i.e., after all players of $S\backslash\{i_s\}$ but before $N\backslash S$, and receives her final utility of $-(s-1)\theta_i - \sum_{j\in S\backslash\{i_s\}} b_j^{i_s} + \sum_{r=s+1}^{n} b_{i_s}^{i_r}$, where $(s-1)\theta_i$ is her waiting cost, $\sum_{j\in S\backslash\{i_s\}} b_j^{i_s}$ is the sum of the bids she paid at stage L-1, and $\sum_{r=s+1}^{n} b_{i_s}^{i_r}$ is the sum of the bids she received from previously rejected proposers $i_{s+1}, \ldots, i_n$. Moreover, stage L-3 is not evoked, but all players in S other than player i_s proceed again from stage L-1 where the set of active players is $S\backslash\{i_s\}$. If the winner i_s decides to sell the position, then she makes a proposal $(\sigma(S), (t_j)_{j\in S})$ consisting of a queue $\sigma(S) \in \Sigma(S)$ and a vector of transfers $(t_j)_{j\in S} \in \mathbf{R}^S$ such that $\sum_{j\in S} t_j \leq 0$. (This offer is additional to the bids paid at stage L-1.) The game continues to stage L-3.

Stage L-3. Approving or disapproving a proposal: The players in S other than player i_s, sequentially, either accept or reject the proposal. If at least one player rejects the proposal, then it is rejected. Otherwise, it is accepted.

(i) If the proposal is rejected, all players in S other than player i_s go back to stage L-1 where the set of active players is $S\backslash\{i_s\}$. Meanwhile, player i_s falls back to her default position: the position after the queue of $S\backslash\{i_s\}$ but before $N\backslash S$. Consequently, player i_s receives her final utility $-(s-1)\theta_i - \sum_{j\in S\backslash\{i_s\}} b_j^{i_s} + \sum_{r=s+1}^{n} b_{i_s}^{i_r}$. Note that she would receive the same utility if she decided in stage L-2 to take up her position.

(ii) If the proposal is accepted, we have to distinguish between two cases, $S = N$ and $S \neq N$. If $S = N$, all players agree with the proposer i_n on her proposal $(\sigma(N), (t_j)_{j\in N})$ and the game ENDS. Each player $j \in N\backslash\{i_n\}$ receives $-(\sigma_j(N) - 1)\theta_j + b_j^{i_n} + t_j$, and player i_n receives $-(\sigma_{i_n}(N) - 1)\theta_{i_n} -$

$\sum_{j \in N\setminus\{i_n\}} b_j^{i_n} + t_{i_n}$ where $t_{i_n} = -\sum_{j \in N\setminus\{i_n\}} t_j$. If $S \neq N$, all players in S agree with the proposer i_s on her proposal $(\sigma(S), (t_j)_{j \in S})$, and the game ENDS. Each player $j \in S\setminus\{i_s\}$ receives $-(\sigma_j(S) - 1)\theta_j + t_j$, and with the bids made by player i_s and all other previously rejected proposers, player j's final utility is $-(\sigma_j(S) - 1)\theta_j + \sum_{k=s}^{n} b_j^{i_k} + t_j$. Player i_s receives $-(\sigma_{i_s}(S) - 1)\theta_{i_s} - \sum_{j \in S\setminus\{i_s\}} b_j^{i_s} + t_{i_s}$, where $t_{i_s} = -\sum_{j \in S\setminus\{i_s\}} t_j$, and by adding the bids player i_s received from the previously rejected proposers, her final utility is $-(\sigma_{i_s}(S) - 1)\theta_{i_s} - \sum_{j \in S\setminus\{i_s\}} b_j^{i_s} + \sum_{k=s+1}^{n} b_{i_s}^{i_k} + t_{i_s}$.

We show that for any $N \in \mathcal{N}$ and any $\theta \in \mathcal{Q}^N$, the last-served mechanism has a unique SPE outcome, which coincides with the payoff vector assigned by the minimal transfer rule. For all $N \in \mathcal{N}$, all $i \in N$, and all $S \subseteq N\setminus\{i\}$, consider an efficient queue σ^* for $S \cup \{i\}$, i.e., $\sigma^* \in Eff(\theta_{S\cup\{i\}})$, and an efficient queue σ^{**} for S, i.e., $\sigma^{**} \in Eff(\theta_S)$. Then, the marginal contribution of player $i \in N$ to coalition $S \subseteq N\setminus\{i\}$ in the optimistic queueing game v_O equals

$$
\begin{aligned}
& v_O(S \cup \{i\}) - v_O(S) \\
&= -\sum_{k \in S\cup\{i\}} (\sigma_k^* - 1)\theta_k - \left(-\sum_{j \in S}(\sigma_j^{**} - 1)\theta_j\right) \\
&= \sum_{j \in S} \sigma_j^{**}\theta_j - \sum_{k \in S\cup\{i\}} \sigma_k^*\theta_k + \theta_i \qquad (9.2) \\
&= \left(\sum_{j \in P_i(\sigma^*)} \sigma_j^*\theta_j + \sum_{j \in F_i(\sigma^*)} (\sigma_j^* - 1)\theta_j\right) - \left(\sum_{j \in P_i(\sigma^*)} \sigma_j^*\theta_j + \sigma_i^*\theta_i + \sum_{j \in F_i(\sigma^*)} \sigma_j^*\theta_j\right) + \theta_i \\
&= (1 - \sigma_i^*)\theta_i - \sum_{j \in F_i(\sigma^*)} \theta_j,
\end{aligned}
$$

where the first equality follows by definition of v_O and the third equality follows from the fact that $\sigma_j^{**} = \sigma_j^*$ for all $j \in P_i(\sigma^*)$ and $\sigma_j^{**} = \sigma_j^* - 1$ for all $j \in F_i(\sigma^*)$. That is, with the joining of player i to coalition S, construction of the efficient queue requires those who have higher waiting costs than θ_i to be served before player i and those who have lower waiting costs than θ_i to be served after player i. Hence, compared to the total cost within coalition S, player i's marginal contribution (in terms of cost) is $-(\sigma_i^* - 1)\theta_i - \sum_{j \in F_i(\sigma^*)} \theta_j$.

We begin with the following lemma.

Lemma 9.2 *For all $N \in \mathcal{N}$, all $\theta \in \mathcal{Q}^N$, all $i \in N$, and all $S \subseteq N\setminus\{i\}$,*

$$v_O(S \cup \{i\}) - v_O(S) \geq -|S|\theta_i.$$

Proof Let $N \in \mathcal{N}$, $\theta \in \mathcal{Q}^N$, $i \in N$, and $S \subseteq N\setminus\{i\}$. Also, let $T \subseteq S$ be the set of players whose unit waiting costs are greater than θ_i and $S\setminus T$ be the set of players whose unit waiting costs are less than or equal to θ_i. That is, $\theta_j > \theta_i \geq \theta_k$ for all $j \in T$ and all $k \in S\setminus T$. In an efficient queue σ^* for $S \cup \{i\}$, player i is served after players in T but before anyone else in the queue σ^*, so that $\sigma_i^* = |T| + 1$.[7] Thus,

$$\begin{aligned} v_O(S \cup \{i\}) - v_O(S) &= (1 - \sigma_i^*)\theta_i - \sum_{k \in F_i(\sigma^*)} \theta_k \\ &= (1 - (|T| + 1))\,\theta_i - \sum_{k \in S \setminus T} \theta_k \\ &= -|T|\theta_i - \sum_{k \in S \setminus T} \theta_k \\ &\geq -|T|\theta_i - (|S| - |T|)\theta_i \\ &= -|S|\theta_i. \end{aligned}$$

□

The inequality in Lemma 9.2 is strict if $\theta_i > \theta_k$ for at least one player $k \in S \setminus \{i\}$. This lemma also offers a desirable implication: it pays for player i to join coalition S and form an efficient queue rather than taking up a position after coalition S since the cost for the coalition $S \cup \{i\}$ is not more than the cost of coalition S plus the cost of player i when she is served in the position after coalition S. Note that the right-hand side of the inequality of lemma is the utility of player i when she takes up the position $\sigma_i = |S| + 1$.

Now we are ready to present our second main result.

Theorem 9.2 *For all $N \in \mathcal{N}$ and all $\theta \in \mathcal{Q}^N$, the last-served mechanism has a unique SPE outcome, which coincides with the payoff vector assigned by the minimal transfer rule $\varphi^M(\theta)$.*

Proof Let $N \in \mathcal{N}$ be such that $N = \{1, \ldots, n\}$ and $\theta \in \mathcal{Q}^N$. The proof proceeds by induction on the number of players n. The induction assumption is that whenever the mechanism is used by n players with a given vector of unit waiting costs, it implements the minimal transfer rule to this queueing problem. It is easy to see that the theorem holds for $n = 1$. We assume that it holds for all $m \leq n - 1$ and show that it is satisfied for n.

First we show that the allocation $\varphi^M(\theta)$ assigned by the minimal transfer rule is an SPE outcome. We explicitly construct an SPE that yields $\varphi^M(\theta)$ as the SPE outcome. Consider the following strategies, which the players would follow in any (sub)game they participate in (we describe it for the whole set of players, N, but similar strategies are followed by any player in $S \subseteq N$ that is called upon to play the game, with S replacing N):

[7] Note that the position of player i may not be $|T| + 1$ if there is a player $j \in S$ with $\theta_j = \theta_i$. Since the choice of an efficient queue has no effect on $v_O(S \cup \{i\}) - v_O(S)$, we can take an efficient queue σ^* with $\sigma_i^* = |T| + 1$.

At stage L-1, each player $i \in N$ announces

$$b_j^i = \left(-(\sigma_j^* - 1)\theta_j + (\sigma_j^* - 1)\frac{\theta_j}{2} - \sum_{k \in F_j(\sigma^*)} \frac{\theta_k}{2}\right)$$
$$- \left(-(\sigma_j^{**} - 1)\theta_j + (\sigma_j^{**} - 1)\frac{\theta_j}{2} - \sum_{k \in F_j(\sigma^{**})} \frac{\theta_k}{2}\right)$$

where $\sigma^* \in \mathit{Eff}(\theta)$ and $\sigma^{**} \in \mathit{Eff}(\theta_{N\backslash\{i_n\}})$ for all $j \in N\backslash\{i\}$.

At stage L-2, the proposer i_n ("winner" of the bidding in stage L-1) adopts the option of selling the position instead of taking it up by herself and makes a proposal (σ^*, t) such that $\sigma^* \in \mathit{Eff}(\theta)$ and $j \in N\backslash\{i_n\}$,

$$t_j = (\sigma_j^* - 1)\theta_j + \left(-(\sigma_j^{**} - 1)\theta_j + (\sigma_j^{**} - 1)\frac{\theta_j}{2} - \sum_{k \in F_j(\sigma^{**})} \frac{\theta_k}{2}\right)$$

where $\sigma^{**} \in \mathit{Eff}(\theta_{N\backslash\{i_n\}})$.

At stage L-3, any player $j \in N\backslash\{i_n\}$ accepts any proposal (σ, t) if $\sigma \in \Sigma(N)$ and

$$t_j \geq (\sigma_j - 1)\theta_j + \left(-(\sigma_j^{**} - 1)\theta_j + (\sigma_j^{**} - 1)\frac{\theta_j}{2} - \sum_{k \in F_j(\sigma^{**})} \frac{\theta_k}{2}\right)$$

where $\sigma^{**} \in \mathit{Eff}(\theta_{N\backslash\{i_n\}})$, and rejects it otherwise.

To verify that the above strategies indeed constitute an SPE and yield the allocation assigned by the minimal transfer rule, we first look at stage L-3. Suppose the proposal of player i_n is rejected. From the induction hypothesis, we know that for the remaining players $N\backslash\{i_n\}$, a corresponding efficient queue is formed in the unique SPE outcome of that game and the resulting utility to every player $j \in N\backslash\{i_n\}$ is $-(\sigma_j^{**} - 1)\theta_j + (\sigma_j^{**} - 1)\frac{\theta_j}{2} - \sum_{k \in F_j(\sigma^{**})} \frac{\theta_k}{2}$, where $\sigma^{**} \in \mathit{Eff}(\theta_{N\backslash\{i_n\}})$, which is the reservation utility for player j when she considers any proposal made by player i_n. If player j accepts the proposal made by player i_n, then she receives the utility $-(\sigma_j - 1)\theta_j + t_j$. Apparently, only when this utility is no less than her reservation utility, she accepts the proposal, which gives rise to $t_j \geq (\sigma_j - 1)\theta_j + \left(-(\sigma_j^{**} - 1)\theta_j + (\sigma_j^{**} - 1)\frac{\theta_j}{2} - \sum_{k \in F_j(\sigma^{**})} \frac{\theta_k}{2}\right)$. Note that with t_j as specified above, at this stage, player j's utility is guaranteed to be no less than her reservation utility regardless of the queue proposed by player i_n.

Now consider stage L-2. Obviously, for any proposed σ, player i_n does not make any player $j \in N\backslash\{i_n\}$ an offer t_j that is strictly higher than $(\sigma_j - 1)\theta_j + \left(-(\sigma_j^{**} - 1)\theta_j + (\sigma_j^{**} - 1)\frac{\theta_j}{2} - \sum_{k \in F_j(\sigma^{**})} \frac{\theta_k}{2}\right)$. In the meantime, player i_n would not lower the offer t_j for any player $j \in N\backslash\{i_n\}$ to be strictly less than $(\sigma_j - 1)\theta_j +$

$\left(-(\sigma_j^{**}-1)\theta_j+(\sigma_j^{**}-1)\frac{\theta_j}{2}-\sum_{k\in F_j(\sigma^{**})}\frac{\theta_k}{2}\right)$ since it would lead her proposal to be rejected and then served at last with the waiting cost of $-(|N|-1)\theta_{i_n}$. We show that player i_n does have an incentive to make such an acceptable proposal. By making this proposal at stage L-2, player i_n receives

$$
\begin{aligned}
&-(\sigma_{i_n}-1)\theta_{i_n}-\sum_{j\in N\setminus\{i_n\}} t_j\\
&=-(\sigma_{i_n}-1)\theta_{i_n}\\
&\quad-\sum_{j\in N\setminus\{i_n\}}\left((\sigma_j-1)\theta_j+\left(-(\sigma_j^{**}-1)\theta_j+(\sigma_j^{**}-1)\frac{\theta_j}{2}-\sum_{k\in F_j(\sigma^{**})}\frac{\theta_k}{2}\right)\right)\\
&=-\sum_{i\in N}(\sigma_i-1)\theta_i-\sum_{j\in N\setminus\{i_n\}}\left(-(\sigma_j^{**}-1)\theta_j\right)-\sum_{j\in N\setminus\{i_n\}}\left((\sigma_j^{**}-1)\frac{\theta_j}{2}-\sum_{k\in F_j(\sigma^{**})}\frac{\theta_k}{2}\right)\\
&=-\sum_{i\in N}(\sigma_i-1)\theta_i-v_O(N\setminus\{i_n\}),
\end{aligned}
$$

where the last equality follows from the definition of v_O and the fact that $\sum_{j\in N\setminus\{i_n\}}\left((\sigma_j^{**}-1)\frac{\theta_j}{2}-\sum_{k\in F_j(\sigma^{**})}\frac{\theta_k}{2}\right)=0$. Moreover, in order for the proposer i_n to maximize what she can achieve, it is obvious that she chooses $\sigma^*\in \mathit{Eff}(\theta)$ because $-\sum_{i\in N}(\sigma_i^*-1)\theta_i\geq -\sum_{i\in N}(\sigma_i-1)\theta_i$ for all $\sigma\in\Sigma(N)\setminus \mathit{Eff}(\theta)$. Hence, player i_n receives $v_O(N)-v_O(N\setminus\{i_n\})\geq-(|N|-1)\theta_{i_n}$, where the inequality follows from Lemma 9.2.

To verify the strategies restricted to stage L-1 constitute an SPE, note that all net bids equal zero, which follows from the fact that $b_j^i=\phi_j(v_O)-\phi_j(v_O|_{N\setminus\{i\}})$ where $v_O|_{N\setminus\{i\}}(S)=v_O(S)$ for all $S\subseteq N\setminus\{i\}$ and by the balanced contributions property of the Shapley value (Myerson 1980; van den Brink and Chun 2012). To show that a change in the bids made by player i cannot increase that player's payoff, we consider the following two cases: First, if player i changes the vector of her bids so that another player becomes the proposer, this does not change her resulting utility, which would still equal that given by the minimal transfer rule (i.e., her allocation assigned by the Shapley value of the optimistic queueing game v_O). Second, if she changes the vector of her bids and following it she is still the proposer with positive probability, it must be that her total bid ($\sum_{j\in N\setminus\{i\}}b_j^i$) does not decline, which again means her payoff cannot improve. That is, any deviation of the bidding strategy of player i specified at stage L-1 cannot improve her payoff. Hence, no player has an incentive to change her bid, showing that the given strategy profile is an SPE.

The proof that any SPE yields the allocation assigned by the minimal transfer rule proceeds by a series of claims.

Claim L-1 At stage L-3, in any SPE, any player $j \in N\backslash\{i_n\}$ accepts any proposal (σ, t) such that $\sigma \in \Sigma(N)$ if

$$t_j \geq (\sigma_j - 1)\theta_j + \left(-(\sigma_j^{**} - 1)\theta_j + (\sigma_j^{**} - 1)\frac{\theta_j}{2} - \sum_{k \in F_j(\sigma^{**})} \frac{\theta_k}{2}\right),$$

and rejects it if

$$t_j < (\sigma_j - 1)\theta_j + \left(-(\sigma_j^{**} - 1)\theta_j + (\sigma_j^{**} - 1)\frac{\theta_j}{2} - \sum_{k \in F_j(\sigma^{**})} \frac{\theta_k}{2}\right)$$

where $\sigma^{**} \in \mathit{Eff}(\theta_{N\backslash\{i_n\}})$.

Proof This claim follows directly from the induction assumption. □

Claim L-2

(i) If $v_O(N) - v_O(N\backslash\{i_n\}) > -(|N| - 1)\theta_{i_n}$, the only SPE of the game that starts at stage L-2 is the following. At stage L-2, player i_n chooses the option of selling the position instead of taking it up by herself and makes a proposal (σ^*, t) such that $\sigma^* \in \mathit{Eff}(\theta)$ and

$$t_j = (\sigma_j^* - 1)\theta_j + \left(-(\sigma_j^{**} - 1)\theta_j + (\sigma_j^{**} - 1)\frac{\theta_j}{2} - \sum_{k \in F_j(\sigma^{**})} \frac{\theta_k}{2}\right)$$

where $\sigma^{**} \in \mathit{Eff}(\theta_{N\backslash\{i_n\}})$ to each player $j \in N\backslash\{i_n\}$. At stage L-3, each player $j \in N\backslash\{i_n\}$ accepts any proposal (σ, t) such that $\sigma \in \Sigma(N)$ if

$$t_j \geq (\sigma_j - 1)\theta_j + \left(-(\sigma_j^{**} - 1)\theta_j + (\sigma_j^{**} - 1)\frac{\theta_j}{2} - \sum_{k \in F_j(\sigma^{**})} \frac{\theta_k}{2}\right)$$

where $\sigma^{**} \in \mathit{Eff}(\theta_{N\backslash\{i_n\}})$ and rejects it otherwise.

(ii) If $v_O(N) - v_O(N\backslash\{i_n\}) = -(|N| - 1)\theta_{i_n}$, there exist other SPEs in addition to the one described above. In fact, any set of the following strategies also constitutes an SPE: at stage L-2, player i_n either takes up the last position by herself or sells the position by making a proposal (σ, t) such that $\sigma \in \Sigma(N)$ and to some player $j \neq i_n$,

$$t_j < (\sigma_j - 1)\theta_j + \left(-(\sigma_j^{**} - 1)\theta_j + (\sigma_j^{**} - 1)\frac{\theta_j}{2} - \sum_{k \in F_j(\sigma^{**})} \frac{\theta_k}{2}\right)$$

where $\sigma^{**} \in Eff(\theta_{N\setminus\{i_n\}})$; and at stage L-3, player j rejects any proposal

$$t_j \leq (\sigma_j - 1)\theta_j + \left(-(\sigma_j^{**} - 1)\theta_j + (\sigma_j^{**} - 1)\frac{\theta_j}{2} - \sum_{k \in F_j(\sigma^{**})} \frac{\theta_k}{2}\right).$$

In any SPE of this subgame, the final payoffs to players i_n and $j \neq i_n$ are $-(\sigma_{i_n}^* - 1)\theta_{i_n} - \sum_{j \in N\setminus\{i_n\}} b_j^{i_n} - \sum_{j \in N\setminus\{i_n\}} t_j$ and $-(\sigma_j^* - 1)\theta_j + b_j^{i_n} + t_j$, respectively, where $t_j = (\sigma_j^* - 1)\theta_j + \left(-(\sigma_j^{**} - 1)\theta_j + (\sigma_j^{**} - 1)\frac{\theta_j}{2} - \sum_{k \in F_j(\sigma^{**})} \frac{\theta_k}{2}\right)$.

Proof For the case of $v_O(N) - v_O(N\setminus\{i_n\}) > -(|N| - 1)\theta_{i_n}$, one can verify the argument by the induction assumption and Lemma 9.2. For the case of $v_O(N) - v_O(N\setminus\{i_n\}) = -(|N| - 1)\theta_{i_n}$, one can obviously see that player i_n would be indifferent between taking up the last position and making an acceptable proposal, with the first option being equivalent to making an unacceptable proposal and then having it be rejected, which would all yield the same payoff to player i_n. □

Claim L-3 In any SPE, $B^i = B^j$ for all $i, j \in N$ and hence $B^i = 0$ for all $i \in N$.

Claim L-4 In any SPE, each player's payoff is the same regardless of who is chosen as the winner at stage L-1.

Proof The proofs of Claims L-3 and L-4 are the same as in Pérez-Castrillo and Wettstein (2001). □

Claim L-5 In any SPE, the final payoff of each player coincides with her payoff assigned by the minimal transfer rule.

Proof Note that if player i is the proposer, her final payoff is

$$-(\sigma_i^* - 1)\theta_i - \sum_{j \in N\setminus\{i\}} b_j^i - \sum_{j \in N\setminus\{i\}} t_j,$$

where $t_j = (\sigma_j^* - 1)\theta_j + \left(-(\sigma_j^{**} - 1)\theta_j + (\sigma_j^{**} - 1)\frac{\theta_j}{2} - \sum_{k \in F_j(\sigma^{**})} \frac{\theta_k}{2}\right)$. And if player $j \neq i$ is the proposer, the final payoff of player i is

$$-(\sigma_i^* - 1)\theta_i + b_i^j + \left((\sigma_i^* - 1)\theta_i + \left(-(\sigma_i^{**} - 1)\theta_i + (\sigma_i^{**} - 1)\frac{\theta_i}{2} - \sum_{k \in F_i(\sigma^{**})} \frac{\theta_k}{2}\right)\right).$$

Therefore, the sum of payoffs to player i over all possible choices of the proposer is

$$\begin{aligned}
& -(\sigma_i^* - 1)\theta_i - \sum_{j\in N\setminus\{i\}} b_j^i \\
& - \sum_{j\in N\setminus\{i\}} \left((\sigma_j^* - 1)\theta_j + \left(-(\sigma_j^{**} - 1)\theta_j + (\sigma_j^{**} - 1)\frac{\theta_j}{2} - \sum_{k\in F_j(\sigma^{**})} \frac{\theta_k}{2} \right) \right) \\
& + \sum_{j\neq i} \left(-(\sigma_i^* - 1)\theta_i + b_i^j \right) \\
& + \sum_{j\neq i} \left((\sigma_i^* - 1)\theta_i + \left(-(\sigma_i^{**} - 1)\theta_i + (\sigma_i^{**} - 1)\frac{\theta_i}{2} - \sum_{k\in F_i(\sigma^{**})} \frac{\theta_k}{2} \right) \right) \\
= \; & -n(\sigma_i^* - 1)\theta_i \\
& - \sum_{j\in N\setminus\{i\}} \left((\sigma_j^* - 1)\theta_j + \left(-(\sigma_j^{**} - 1)\theta_j + (\sigma_j^{**} - 1)\frac{\theta_j}{2} - \sum_{k\in F_j(\sigma^{**})} \frac{\theta_k}{2} \right) \right) \\
& + (n-1)(\sigma_i^* - 1)\theta_i + \sum_{j\neq i} \left(-(\sigma_i^{**} - 1)\theta_i + (\sigma_i^{**} - 1)\frac{\theta_i}{2} - \sum_{k\in F_i(\sigma^{**})} \frac{\theta_k}{2} \right) \\
= \; & -(\sigma_i^* - 1)\theta_i - \sum_{j\in N\setminus\{i\}} (\sigma_j^* - 1)\theta_j - \sum_{j\in N\setminus\{i\}} \left(-(\sigma_j^{**} - 1)\theta_j \right) \\
& + \sum_{j\in N\setminus\{i\}} \left(-(\sigma_i^{**} - 1)\theta_i + (\sigma_i^{**} - 1)\frac{\theta_i}{2} - \sum_{k\in F_i(\sigma^{**})} \frac{\theta_k}{2} \right) \\
= \; & -(\sigma_i^* - 1)\theta_i - \sum_{k\in F_i(\sigma^*)} \theta_k + \sum_{j\in N\setminus\{i\}} \left(-(\sigma_i^{**} - 1)\frac{\theta_i}{2} - \sum_{k\in F_i(\sigma^{**})} \frac{\theta_k}{2} \right) \\
= \; & -\sigma_i^*\theta_i + \theta_i - \sum_{k\in F_i(\sigma^*)} \theta_k - \sum_{j\in N\setminus\{i\}} \frac{\sigma_i^{**}\theta_i}{2} + (n-1)\frac{\theta_i}{2} - \sum_{j\in N\setminus\{i\}} \sum_{k\in F_i(\sigma^{**})} \frac{\theta_k}{2} \\
= \; & -\sigma_i^*\theta_i - \sum_{k\in F_i(\sigma^*)} \theta_k - \sum_{j\in N\setminus\{i\}} \frac{\sigma_i^{**}\theta_i}{2} + (n+1)\frac{\theta_i}{2} - \sum_{j\in N\setminus\{i\}} \sum_{k\in F_i(\sigma^{**})} \frac{\theta_k}{2} \\
= \; & -\sigma_i^*\theta_i - \sum_{k\in F_i(\sigma^*)} \theta_k - \frac{1}{2}\left(|F_i(\sigma^*)|\sigma_i^*\theta_i + |P_i(\sigma^*)|(\sigma_i^* - 1)\theta_i \right) + (n+1)\frac{\theta_i}{2} \\
& - \sum_{j\in N\setminus\{i\}} \sum_{k\in F_i(\sigma^{**})} \frac{\theta_k}{2} \\
= \; & -\frac{n+1}{2}(\sigma_i^* - 1)\theta_i + \frac{1}{2}|P_i(\sigma^*)|\theta_i - \sum_{k\in F_i(\sigma^*)} \theta_k - \frac{1}{2}(|P_i(\sigma^*)| + |F_i(\sigma^*)| - 1) \sum_{k\in F_i(\sigma^*)} \theta_k
\end{aligned}$$

$$= -\frac{n+1}{2}(\sigma_i^* - 1)\theta_i + \frac{1}{2}(\sigma_i^* - 1)\theta_i - \frac{n}{2}\sum_{k \in F_i(\sigma^*)} \theta_k$$

$$= -\frac{n}{2}(\sigma_i^* - 1)\theta_i - \sum_{k \in F_i(\sigma^*)} \frac{n\theta_k}{2}$$

$$= n\left(-(\sigma_i^* - 1)\theta_i + (\sigma_i^* - 1)\frac{\theta_i}{2} - \sum_{k \in F_i(\sigma^*)} \frac{\theta_k}{2}\right)$$

$$= n\varphi_i^M(\theta).$$

Together with Claim L-4, we conclude that in any SPE, for any player $i \in N$, her final payoff is $\varphi_i^M(\theta)$. □

Remark 9.1 As shown in Sect. 3.5, the Shapley value and the nucleolus (or the prenucleolus) coincide in queueing problems. Together with Theorems 9.1 and 9.2, our games can implement the nucleolus (or the prenucleolus) of the corresponding games.

Here we discuss implications of the two mechanisms implementing the maximal and the minimal transfer rules, respectively. In both mechanisms, the players have the same strategies. However, the two mechanisms assign different positions to a winner who decides to take up the position in stage 2, or whose proposal is rejected in stage 3, of a certain round.[8] In the first-served mechanism, this player gets the first position (after the already rejected players), while in the last-served mechanism, this player gets the last position (in front of the already rejected players). At first sight, the first-served mechanism seems to be advantageous for the proposer since she takes up the best available position (either in stage F-2 or if her proposal is rejected in stage F-3). Moreover, the proposer seems to have an incentive of making her proposal be rejected since it gets the best position if her proposal is accepted. On the other hand, in the last-served mechanism, the proposer seems to have an incentive of making her proposal be accepted since it gets the worst position if her proposal is rejected.

The "value" of being the proposer, and therefore the bids made in stage 1, depends on the position that is at stake for the proposer. As it turns out, in SPE, being the proposer in the first-served mechanism is so attractive that the bids to become the proposer are so high that it eventually leads to a combination of bids (in stage F-1) and offers (in stage F-2) such that the SPE outcome yields the utility payoffs assigned by the maximal transfer rule which asks relatively large compensations from the players served earlier to the players served later. On the other hand, in SPE, being the proposer in the last-served mechanism is so unattractive that the bids to become the proposer are so low (in fact, the players want to be paid to become the

[8] Note that the possibility of taking up the position and leaving the game is not a part of the Pérez-Castrillo and Wettstein's (2001) mechanism which implements the Shapley value for TU games.

proposer) that it eventually leads to a combination of bids (in stage L-1) and offers (in stage L-2) such that the SPE outcome yields the utility payoffs assigned by the minimal transfer rule which asks relatively small compensations from the players served earlier to the players served later. Requiring a rejected player to be served last leads to an equilibrium outcome corresponding to a rule that yields smaller transfers that agents with higher waiting costs have to make to agents with lower waiting costs in compensation for taking up a position in front of the queue.

Our result provides a link between a pessimistic treatment of a proposer and an optimistic treatment of a coalition aligning together for the minimal transfer rule. The minimal transfer rule can be obtained by the last-served mechanism that assigns the last position to the proposer, a pessimistic treatment of a proposer. This rule can also be obtained under the optimistic assumption that a coalition is served before the rest of players (see Sect. 3.3 for details). Similarly, the first-served mechanism provides a link between an optimistic treatment of a proposer and a pessimistic treatment of a coalition, which in either way leads to the same maximal transfer rule.

9.4 The Hybrid Mechanism and the Average Transfer Rule

The axiomatic foundations of the maximal and the minimal transfer rules indicate that the two rules have desirable properties and are complementary to each other. Next we introduce an average transfer rule that takes the average of these two rules, which can be seen as a compromised choice between the optimistic and the pessimistic perspectives. To construct a strategic mechanism for the average transfer rule, one may combine the two mechanisms by setting up a lottery device in the beginning such that there will be an equal probability to play the first-served and the last-served mechanisms. Each player's expected payoff from playing this mega-game equals the average of the allocations generated by the maximal and the minimal transfer rules. However, this is not a natural implementation mechanism because players' payoffs are generated exogenously and the coincidence only happens in expectation rather than in actual terms. To deal with this difficulty, we offer another mechanism implementing the average transfer rule.

Hybrid Mechanism The mechanism is the same as the previous two mechanisms, except when the set of active players is N.

Stage H-1. Bidding for the proposer: It is the same as stage 1 of the previous two mechanisms, except that the winner in this mechanism has an equal chance to be served in the first and the last positions (e.g., by a fair lottery device like flipping a coin), instead of being served for sure either in the first position or in the last position.

Stage H-2. Taking the chance or making a proposal: The winner i_n decides to either take the chance by herself or make a proposal. If player i_n takes the chance by herself, a lottery device generates the actual state for player i_n to be located first or last. Then, the mechanism proceeds along this realized position for all the other players, with no lottery device anymore. If player i_n is located first by the lottery device, then the mechanism proceeds as in the first-served mechanism for the other players, and if she is located last by the lottery device, then the mechanism proceeds as in the last-served mechanism for the other players. On the other hand, if player i_n decides to make a proposal, then the proposal consists of a queue $\sigma(N) \in \Sigma(N)$ and a vector of transfers $(t_j)_{j \in N} \in \mathbf{R}^N$ such that $\sum_{j \in N} t_j \leq 0$. The game continues to stage H-3.

Stage H-3. Approving or disapproving a proposal: The players in N other than player i_n, sequentially, either accept or reject the proposal. If at least one player rejects the proposal, then it is rejected. Otherwise, it is accepted. If the proposal is rejected, then the lottery device reveals which position, the first or the last, the rejected proposer i_n actually takes. For all the other players $N \backslash \{i_n\}$, the mechanism will proceed according to the realized position on player i_n, and there will be no lottery device anymore. That is, if player i_n takes up the first position, then the other players will play the first-served mechanism. On the contrary, if player i_n takes up the last position, then the other players will play the last-served mechanism.

To sum up, the only difference takes place when the active set of players is N, where a proposer has an equal probability to be served in the first position and the last position. Once the position is revealed, then the mechanism proceeds along this line for the remaining players.

Proposition 9.1 *For any $N \in \mathcal{N}$ and any $\theta \in \mathcal{Q}^N$, the hybrid mechanism has a unique SPE outcome, which coincides with the payoff vector assigned by the average transfer rule, i.e., $\frac{1}{2}\left(\varphi^M(\theta) + \varphi^C(\theta)\right)$.*

Proof Since the proof can be constructed along the same line as in Theorems 9.1 and 9.2, we only provide the sketch. From Theorem 9.1, if the first-served mechanism is played by $N \backslash \{i\}$, then the unique SPE outcome for this subgame is $\varphi^C(\theta_{N \backslash \{i_n\}})$. Similarly, if the last-served mechanism will be played, then the outcome is $\varphi^M(\theta_{N \backslash \{i_n\}})$. Therefore, in SPE, player i_n will make a proposal such that each player $j \in N \backslash \{j\}$ obtains $\frac{1}{2}\left(\varphi_j^C(\theta_{N \backslash \{i_n\}}) + \varphi_j^M(\theta_{N \backslash \{i_n\}})\right)$, which is player j's expected payoff if the proposal by player i_n is rejected. Due to Lemmas 9.1 and 9.2, one can obviously see that player i_n has an incentive to make such an acceptable proposal instead of being rejected and receiving the expected payoff $\frac{1}{2}\left(0 - (n-1)\theta_{i_n}\right)$. □

9.5 Independence of the Ordering of the Rejected Players

Now we discuss a robustness property of the two mechanisms implementing the minimal and the maximal transfer rules. Recall that the last-served mechanism requires a rejected proposer to be served after the others. If the proposal of player i_n, as the proposer for the set of active players N, is rejected, then she is placed at the last position of the entire queue of N, no matter how players in $N\backslash\{i_n\}$ are arranged. If the proposal of player i_{n-1}, as the proposer for the set of active players $N\backslash\{i_n\}$, is rejected, then she takes up the $(|N|-1)$th position but before player i_n and so on. The most recently rejected player will be served after the remaining set of active players, but before the previously rejected players. One may consider an alternative design of the mechanism which assigns to the first rejected proposer the first position of all the rejected players and to the most recently rejected player the last position of all the rejected players. The following proposition confirms that such a mechanism would still implement the minimal transfer rule.

Proposition 9.2 *The ordering of the rejected players in the last-served mechanism, so long as they are served after the set of active players S, plays no role in implementing the minimal transfer rule of the queueing problem in SPE.*

Proof Any alternative ordering different from the one specified in the original last-served mechanism would mean that the most recently rejected player cannot do better. That is, player i_s, if rejected, takes the sth position or after. Taking the sth position, the same as in the original last-served mechanism, gives her no incentive to make an unacceptable offer when she makes a proposal. Since a later position can only make her worse off when being rejected, she makes an acceptable offer in SPE. Therefore, even if a rejected player is placed at the end of all the rejected players, she would see that all others have no incentive to make their offers be rejected and this ordering gives the same result as in the original last-served mechanism. □

Similarly, we can construct an alternative mechanism to implement the maximal transfer rule. In this mechanism, when a proposer is rejected, her final position in the queue of N is pending, up to the moment when the finally rejected proposer is settled. Suppose that coalition $S\backslash\{i_s\}$ have made an agreement. The last rejected proposer i_s takes up the position immediately before $S\backslash\{i_s\}$ but after $N\backslash S$, whereas all players in $N\backslash S$ can form any ordering.

Proposition 9.3 *The ordering of the rejected players in the first-served mechanism, so long as they are served before the set of active players S and the last rejected player takes the position immediately before S, plays no role in implementing the maximal transfer rule of a queueing problem in SPE.*

Proof A player may wish her followers to become rejected players so that she might be better offered by being placed at an earlier position. However, in a two-player subgame, the proposer would not make an unacceptable offer. So the previous proposer, for a three-player subgame, can foresee this and realize that she would have to be the finally rejected proposer, if she were to make a proposal being rejected, which implies that she is placed immediately before the other two players, but not further front. Therefore, she would not make an unacceptable offer, either. Backward induction leads the first proposer to make an acceptable offer, too. □

References

Chun, Y. (2006). A pessimistic approach to the queueing problem. *Mathematical Social Sciences, 51*, 171–181.

Gul, F. (1989). Bargaining foundations of Shapely value. *Econometrica, 57*, 81–95.

Hart, S., & Mas-Colell, A. (1996). Bargaining and value. *Econometrica, 64*, 357–380.

Ju, Y. (2013). Efficiency and compromise: A bid-offer counteroffer mechanism with two players. *International Journal of Game Theory, 42*, 501–520.

Ju, Y., & Wettstein, D. (2009). Implementing cooperative solution concepts: A generalized bidding approach. *Economic Theory, 39*, 307–330.

Ju, Y., Chun, Y., & van den Brink, R. (2014a). Auctioning and selling positions: A non-cooperative approach to queueing conflicts. Tinbergen Institute Discussion Paper TI 2014-016/II, The Netherlands.

Ju, Y., Chun, Y., & van den Brink, R. (2014b). Auctioning and selling positions: A non-cooperative approach to queueing conflicts. *Journal of Economic Theory, 153*, 33–45.

Maniquet, F. (2003). A characterization of the Shapley value in queueing problems. *Journal of Economic Theory, 109*, 90–103.

Moulin, H. (1981). Implementing just and efficient decision making. *Journal of Public Economics, 16*, 193–213.

Myerson, R. B. (1980). Conference structures and fair allocation rules. *International Journal of Game Theory, 9*, 169–182.

Pérez-Castrillo, D., & Wettstein, D. (2001). Bidding for the surplus: A non-cooperative approach to the Shapley value. *Journal of Economic Theory, 100*, 274–294.

Shapley, L. S. (1953). A Value for n-person Games. In H. W. Kuhn & A. W. Tucker (Eds.), *Contributions to the theory of games II*. Annals of mathematics studies (Vol. 28, pp. 307–317). Princeton: Princeton University Press.

van den Brink, R., & Chun, Y. (2012). Balanced consistency and balanced cost reduction for sequencing problems. *Social Choice and Welfare, 38*, 519–529.

Chapter 10
Queueing Problems with Two Parallel Servers

10.1 Introduction

In this chapter, we generalize the queueing problem with one server (or one-server queueing problem) by assuming the facility has two parallel servers so that two agents can be served at the same time. We introduce two rules for the queueing problem with two parallel servers (or two-server queueing problem), which we also call the minimal transfer rule and the maximal transfer rule, and discuss their properties. The minimal transfer rule is proposed by Maniquet (2003) for the one-server queueing problem in which the facility serves one agent at a time. For each queueing problem, this rule assigns the same payoff as the Shapley value (Shapley 1953) of a TU (transferable utility) game associated with the one-server queueing problem which is obtained by defining the worth of each coalition to be the minimal total waiting cost incurred by its members, assuming optimistically that they are served before the non-coalitional members. On the other hand, the maximal transfer rule is proposed by Chun (2006) for the one-server queueing problem. For each problem, the rule also assigns the same payoff as the Shapley value of a TU game associated with the one-server queueing problem, but this time the worth of each coalition is pessimistically defined; the coalitional members are served after the non-coalitional members (see Chap. 3 for details).

For the two-server queueing problem, the two rules also correspond to the Shapley value of TU games obtained by making two different assumptions on the worth of a coalition. What complicates the analysis is that we need to allow transfers between two agents served at the same time. Thus, our results cannot be obtained by a simple adaptation of the results for the one-server queueing problem. Even though two agents are served at the same time, their compensations may be different: the difference in the transfers can be interpreted as being due to the agent with the smaller unit waiting cost having to receive a compensation from the agent with the larger unit waiting cost. If an agent in a preceding position leaves, then the one with the larger unit waiting cost moves up by one position, saving one unit of her waiting

© Springer International Publishing Switzerland 2016

Y. Chun, *Fair Queueing*, Studies in Choice and Welfare,
DOI 10.1007/978-3-319-33771-5_10

cost. This possibility requires that a monetary transfer be imposed on agents in the same position as well as on agents in different positions.

The chapter is organized as follows.[1] In Sect. 10.2, we introduces the two-server queueing problem and presents basic concepts. In Sect. 10.3, we introduce the optimistic two-server queueing game and show that the Shapley value applied to this game gives the same payoff as the minimal transfer rule. In Sect. 10.4, we introduce the pessimistic two-server queueing game and show that the Shapley value applied to this game gives the same payoff as the maximal transfer rule. Finally, in Sect. 10.5, we investigate the consequences of applying other cooperative game theoretic solutions to the two-server queueing games and the existence of rules satisfying *strategyproofness*. Also, we discuss a possible extension of the two-server queueing problem to the multiple-server queueing problem.

10.2 Two-Server Queueing Problems

We generalize the one-server queueing problem to accommodate two servers. Let $I \equiv \{1, 2, \cdots\}$ be an (infinite) universe of "potential" agents and $\mathcal{N}$ be the family of nonempty finite subsets of I. Each agent $i \in I$ is characterized by her unit waiting cost, $\theta_i \geq 0$. The facility can serve two agents at a time and each agent needs one unit of time for her job to be processed. Given $N \in \mathcal{N}$, each agent $i \in N$ is assigned a position $g_i \in \mathbf{N}$ in the queue and a positive or negative transfer $t_i \in \mathbf{R}$. An agent who is served first incurs no waiting cost. If agent $i \in N$ is served in the g_ith position, his total waiting cost is $(g_i - 1)\theta_i$. Each agent $i \in N$ has a preference over assignments that can be represented by a quasi-linear utility function: her utility from the assignment (g_i, t_i) is given by $u(g_i, t_i; \theta_i) = -(g_i - 1)\theta_i + t_i$.

Given $N \in \mathcal{N}$, a queueing problem with ℓ parallel servers or an *ℓ-server queueing problem* is defined as a list $q = (\theta; \ell)$ where $\theta \in \mathbf{R}^N_+$ is the vector of their unit waiting costs and ℓ is the number of servers. From now on, since we only analyze the two-server case in this chapter, we drop ℓ from the notation and denote the two-server queueing problem by θ. Let $\mathcal{T}^N$ be the class of all two-server queueing problems for N and $\mathcal{T} = \bigcup \mathcal{T}^N$. An *allocation* for $\theta \in \mathcal{T}^N$ is a pair (g, t), where for each $i \in N$, g_i is the position assigned to agent i and t_i is her monetary transfer. An allocation is *feasible* if at most two agents are assigned to each position and the sum of transfers is not positive. Thus, the set of feasible allocations $Z(\theta)$ consists of all pairs (g, t) such that for each $i \in N$, $|\{j \in N | g_j = g_i, j \neq i\}| \leq 1$ and $\sum_{j \in N} t_j \leq 0$.

As in the one-server queueing problem, an allocation is (Pareto) *efficient* if it is feasible, and there is no other feasible allocation that each agent finds at least as desirable and at least one agent finds more desirable. An efficient queue is obtained

[1]This chapter is based mainly on Chun and Heo (2008). Excerpts from that article are reprinted with kind permission of Wiley.

by minimizing total waiting cost. That is, given $N \in \mathcal{N}$ and $\theta \in \mathcal{T}^N$, a queue g is *queue-efficient* if for all g', $\sum_{i \in N}(g_i - 1)\theta_i \leq \sum_{i \in N}(g'_i - 1)\theta_i$. The efficient queue of a problem does not depend on the transfers. Moreover, it is unique except for agents with equal waiting costs. These agents have to be served consecutively but in any order. Let $Eff(\theta)$ be the set of all efficient queues for $\theta \in \mathcal{T}^N$.

For all $N \in \mathcal{N}$ and all $\theta \in \mathcal{T}^N$, an allocation $(g, t) \in Z(\theta)$ is *budget balanced* if $\sum_{i \in N} t_i = 0$. An allocation rule, or simply a *rule*, is a mapping $\varphi : \mathcal{T} \to \sum_{N \in \mathcal{N}} Z(\theta)$, which associates with all $N \in \mathcal{N}$ and all $\theta \in \mathcal{T}^N$ a nonempty subset $\varphi(\theta)$ of feasible allocations. The pair $\varphi_i(\theta) = (g_i, t_i)$ represents the position of agent i in the queue and her transfer in θ under φ.

To facilitate our analysis, we assume that agents are indexed by the non-increasing order of their waiting costs; the agent indexed 1 has the largest cost, the agent indexed 2 has the second largest cost, and so on. This indexing is uniquely defined except when agents have equal waiting costs. Agents with equal unit waiting costs have to be indexed consecutively. For all $N \in \mathcal{N}$ and all $\theta \in \mathcal{T}^N$, let $D(\theta)$ be the set of all possible index assignments. For all $d \in D(\theta)$ and all $i \in N$, the efficient queue g is defined as

$$g_i = \lceil \frac{d_i}{2} \rceil = \begin{cases} \frac{d_i}{2} & \text{if } d_i \text{ is even,} \\ \frac{d_i+1}{2} & \text{if } d_i \text{ is odd.} \end{cases}$$

For all $N \in \mathcal{N}$, all $\theta \in \mathcal{T}^N$, all $d \in D(\theta)$, and all $i \in N$, let $P_i(d)$ be the set of agents with smaller indices than agent i and $F_i(d)$ the set of agents with larger indices than agent i.

10.3 An Optimistic Approach and the Minimal Transfer Rule

As in Chap. 3, we solve the two-server queueing problem by using cooperative game theoretic approach. First, we define the worth of each coalition $S \subseteq N$ to be the minimal waiting cost incurred by its members under the optimistic assumption that they are served before the non-coalitional members (Maniquet 2003). That is, for all $S \subseteq N$, its worth $v_O(S)$ is defined by setting

$$v_O(S) = -\sum_{i \in S}(g_i^S - 1)\theta_i,$$

where $\theta_S = (\theta_i)_{i \in S}$ and $g^S \in Eff(\theta_S)$.

Now we define the minimal transfer rule. This name is justified in Remarks 10.2 and 10.3. First, this rule chooses an efficient queue. Second, it assigns to agent i a two-part transfer. One part can be interpreted as a compensation to agent i, the other is a contribution from her. The compensation to agent i is the sum of the position number of all preceding agents divided by her index and multiplied by her unit cost. This compensation to agent i is collected equally from agents with smaller indices.

Thus, the contribution from an agent is the sum of all the amounts contributed by her to all the other agents with larger indices.

Minimal transfer rule, φ^M: For all $N \in \mathcal{N}$ and all $\theta \in \mathcal{T}^N$,

$$\varphi^M(\theta) = \{(g^M, t^M) \in Z(\theta) \mid \forall d \in D(\theta) \text{ and } \forall i \in N,\ g_i^M = \lceil \tfrac{d_i}{2} \rceil \text{ and}$$
$$t_i^M = \frac{\sum_{g_j^M < g_i^M} g_j^M \cdot 2}{d_i} \cdot \theta_i - \sum_{k \in F_i(d)} \left(\frac{1}{d_k - 1} \cdot \frac{\sum_{g_j^M < g_k^M} g_j^M \cdot 2}{d_k} \cdot \theta_k\right)\}.$$

Remark 10.1 Alternatively, the transfer can be expressed as follows. For all $i \in N$,

$$t_i^M = \frac{g_i^M(g_i^M - 1)}{d_i}\theta_i - \sum_{k \in F_i(d)} \frac{1}{d_k - 1} \frac{g_k^M(g_k^M - 1)}{d_k}\theta_k.$$

Since this expression is somewhat easier to manipulate, we will use it in the proofs.

Remark 10.2 If there is only one server, agent i's position g_i can be replaced by index d_i, and 2 in the numerator of the compensation term changed to 1. Altogether, the definition becomes

$$\begin{aligned} t_i^M &= \frac{\sum_{d_j < d_i} d_j \cdot 1}{d_i} \cdot \theta_i - \sum_{k \in N: d_k > d_i} \left(\frac{1}{d_k - 1} \cdot \frac{\sum_{d_j < d_k} d_j \cdot 1}{d_k} \cdot \theta_k\right) \\ &= \frac{d_i - 1}{2}\theta_i - \textstyle\sum_{k \in N: d_k > d_i} \frac{1}{2}\theta_k, \end{aligned}$$

which coincides with the minimal transfer rule for the one-server queueing problem. Moreover, as discussed in Remark 10.3, the minimal transfer rule for the two-server queueing problem can be characterized by imposing the same axioms as in the one-server queueing problem.

Here is an example showing how the minimal transfer rule of the two-server queueing problem is calculated.

Example 10.1 (The Minimal Transfer Rule) Let $N = \{1, 2, 3, 4, 5, 6\}$ and $\theta = (\theta_1, \theta_2, \theta_3, \theta_4, \theta_5, \theta_6)$ be such that $\theta_1 > \theta_2 > \theta_3 > \theta_4 > \theta_5 > \theta_6$. By *queue-efficiency*, $g^M = (1, 1, 2, 2, 3, 3)$ and $d = (1, 2, 3, 4, 5, 6)$. The transfers assigned by the minimal transfer rule are

$$\begin{cases} t_1^M = -\frac{1}{2}\frac{2}{3}\theta_3 - \frac{1}{3}\frac{1}{2}\theta_4 - \frac{1}{4}\frac{6}{5}\theta_5 - \frac{1}{5}\theta_6 \\ t_2^M = -\frac{1}{2}\frac{2}{3}\theta_3 - \frac{1}{3}\frac{1}{2}\theta_4 - \frac{1}{4}\frac{6}{5}\theta_5 - \frac{1}{5}\theta_6 \\ t_3^M = \frac{1 \cdot 2}{3}\theta_3 - \frac{1}{3}\frac{1}{2}\theta_4 - \frac{1}{4}\frac{6}{5}\theta_5 - \frac{1}{5}\theta_6 \\ t_4^M = \frac{1 \cdot 2}{4}\theta_4 - \frac{1}{4}\frac{6}{5}\theta_5 - \frac{1}{5}\theta_6 \\ t_5^M = \frac{1 \cdot 2 + 2 \cdot 2}{5}\theta_5 - \frac{1}{5}\theta_6 \\ t_6^M = \frac{1 \cdot 2 + 2 \cdot 2}{6}\theta_6. \end{cases}$$

Agents 3 and 4 occupy the same position, but the minimal transfer rule assigns different amounts of transfers to them. Agent 3's transfer contains the term $-\frac{1}{6}\theta_4$, which can be regarded as a contribution from agent 3 to agent 4. If either agent 1 or agent 2 leaves the queue without being served, then one agent in the second position moves up to the first position. By *queue-efficiency* of the minimal transfer rule, it has to be an agent with the larger unit waiting cost, agent 3. This possibility explains why the transfers of agents 3 and 4 are different even though they occupy the same position.

Next is an example showing that the Shapely value applied to the optimistic two-server queueing game results in the same payoff vector as the minimal transfer rule.

Example 10.2 (Example 10.1 Continued) Let N and θ be defined as in Example 10.1. The Shapley value assigns each agent the following payoff: $Sh(v_O) = (-\frac{1}{3}\theta_3-\frac{1}{6}\theta_4-\frac{3}{10}\theta_5-\frac{1}{5}\theta_6, -\frac{1}{3}\theta_3-\frac{1}{6}\theta_4-\frac{3}{10}\theta_5-\frac{1}{5}\theta_6, -\frac{1}{3}\theta_3-\frac{1}{6}\theta_4-\frac{3}{10}\theta_5-\frac{1}{5}\theta_6, -\frac{1}{2}\theta_4-\frac{3}{10}\theta_5-\frac{1}{5}\theta_6, -\frac{4}{5}\theta_5-\frac{1}{5}\theta_6, -\theta_6)$.

Since for all $i \in N$, $t_i = u_i + (g_i - 1)\theta_i$, the corresponding transfer is $t = (-\frac{1}{3}\theta_3-\frac{1}{6}\theta_4-\frac{3}{10}\theta_5-\frac{1}{5}\theta_6, -\frac{1}{3}\theta_3-\frac{1}{6}\theta_4-\frac{3}{10}\theta_5-\frac{1}{5}\theta_6, \frac{2}{3}\theta_3-\frac{1}{6}\theta_4-\frac{3}{10}\theta_5-\frac{1}{5}\theta_6, \frac{1}{2}\theta_4-\frac{3}{10}\theta_5-\frac{1}{5}\theta_6, \frac{6}{5}\theta_5-\frac{1}{5}\theta_6, \theta_6)$, which coincides with the transfer assigned by the minimal transfer rule in Example 10.1.

To show the relation between the minimal transfer rule and the Shapley value, we first discuss how the dividend can be calculated for optimistic two-server queueing games. It is well-known that a TU-game v can be written as a linear combination of unanimity games, that is, $v = \sum_{T\subseteq N} \lambda_v(T)u_T$, where the unanimity game u_T on N is given by $u_T(S) = 1$ if $T \subseteq S$ and $u_T = 0$ otherwise. For all $S \subseteq N$, its dividend $\lambda_v(S)$ is defined as follows: if $|S| = 1$, then $\lambda_v(S) = v(S)$, and if $|S| > 1$, $\lambda_v(S) = v(S) - \sum_{T\subset S,\, T\neq S} \lambda_v(T)$.

Lemma 10.1 *For all $N \in \mathcal{N}$ and $\theta \in \mathcal{T}^N$, the unanimity coefficient of $S \subseteq N$ is defined as*

$$\lambda_{v_O}(S) = \begin{cases} 0 & \textit{if } |S| = 1 \textit{ or } 2, \\ -(-2)^{|S|-3} \min_{i\in S}\theta_i & \textit{if } |S| \geq 3. \end{cases}$$

Proof If $|S| = 1$ or 2, the conclusion is obtained trivially from $v_O(S) = 0$. If $|S| = 3$, $\lambda_{v_O}(S) = v_O(S) - \sum_{T\subset S,\, T\neq S}\lambda_{v_O}(T) = v_O(S) = -\min_{i\in S}\theta_i = -(-2)^{3-3}\min_{i\in S}\theta_i$, as desired. Now we proceed by induction.

Suppose that the conclusion holds for all $S \subseteq N$ such that $|S| \leq s - 1$. (In particular, we have shown that the conclusion holds for $|S| - 1 \leq 2$.) We need to show that the conclusion holds for $|S| = s$. Without loss of generality, we assume that $S = \{1, 2, \ldots, s\}$, $\theta_1 \geq \theta_2 \geq \cdots \geq \theta_s$, and, for all $i \in S$, $d_i = i$. Thus, for all $i \in S$, if i is odd, $g_i = \frac{i+1}{2}$, and if i is even, $g_i = \frac{i}{2}$. We use the binomial theorem $(a+b)^n = \sum_{i=0}^{n}\binom{n}{i}a^i b^{n-i}$ after replacing a with -2 and b with 1. First, we rewrite

$\lambda_{v_O}(S)$:

$$\begin{aligned}
\lambda_{v_O}(S) &= v_O(S) - \sum_{T \subset S,\, T \neq S} \lambda_{v_O}(T) \\
&= v_O(S) + \sum_{T \subset S,\, T \neq S,\, |T| \geq 3} (-2)^{|T|-3} \min_{i \in T} \theta_i \\
&= v_O(S) + \sum_{i=3}^{s-1} \left\{ \sum_{j=3}^{i} \binom{i-1}{j-1} (-2)^{j-3} \right\} \theta_i + \sum_{j=3}^{s-1} \binom{s-1}{j-1} (-2)^{j-3} \theta_s .
\end{aligned}$$

By the binomial theorem and the definition of g_i, the second term is rewritten as

$$\begin{aligned}
\sum_{i=3}^{s-1} \left\{ \sum_{j=3}^{i} \binom{i-1}{j-1} (-2)^{j-3} \right\} \theta_i &= \sum_{i=3}^{s-1} \left\{ \sum_{j=2}^{i-1} \binom{i-1}{j} (-2)^{j-2} \right\} \theta_i \\
&= \sum_{i=3}^{s-1} \frac{1}{4} \left\{ \sum_{j=2}^{i-1} \binom{i-1}{j} (-2)^{j} (1)^{i-1-j} \right\} \theta_i \\
&= \sum_{i=3}^{s-1} \frac{1}{4} \left[\left\{ \sum_{j=0}^{i-1} \binom{i-1}{j} (-2)^{j} (1)^{i-1-j} \right\} - 1 + 2(i-1) \right] \theta_i \\
&= \sum_{i=3}^{s-1} \frac{1}{4} \left\{ (-2+1)^{i-1} - 1 + 2(i-1) \right\} \theta_i \\
&= \sum_{i=3}^{s-1} \frac{(-1)^{i-1} - 1 + 2(i-1)}{4} \theta_i \\
&= \sum_{i=3}^{s-1} (g_i - 1) \theta_i . \qquad (10.1)
\end{aligned}$$

Similarly, the third term is rewritten as

$$\begin{aligned}
\sum_{j=3}^{s-1} \binom{s-1}{j-1} (-2)^{j-3} \theta_s &= \sum_{j=2}^{s-2} \binom{s-1}{j} (-2)^{j-2} \theta_s \\
&= \sum_{j=2}^{s-1} \binom{s-1}{j} (-2)^{j-2} \theta_s - (-2)^{s-3} \theta_s \\
&= \frac{1}{4} \sum_{j=2}^{s-1} \binom{s-1}{j} (-2)^{j} \theta_s - (-2)^{s-3} \theta_s
\end{aligned}$$

$$
\begin{aligned}
&= \frac{1}{4}\left\{\sum_{j=0}^{s-1}\binom{s-1}{j}(-2)^j - 1 + 2(s-1)\right\}\theta_s - (-2)^{s-3}\theta_s \\
&= \frac{1}{4}\left\{(-2+1)^{s-1} - 1 + 2(s-1)\right\}\theta_s - (-2)^{s-3}\theta_s \\
&= \frac{(-1)^{s-1} - 1 + 2(s-1)}{4}\theta_s - (-2)^{s-3}\theta_s \\
&= (g_s - 1)\theta_s - (-2)^{s-3}\theta_s.
\end{aligned}
$$

Since $v_O(S) = -\sum_{i=3}^{s}(g_i - 1)\theta_i$,

$$
\begin{aligned}
\lambda_{v_O}(S) &= v_O(S) + \sum_{i=3}^{s-1}\left\{\sum_{j=3}^{i}\binom{i-1}{j-1}(-2)^{j-3}\right\}\theta_i + \sum_{j=3}^{s-1}\binom{s-1}{j-1}(-2)^{j-3}\theta_s \\
&= -\sum_{i=3}^{s}(g_i - 1)\theta_i + \sum_{i=3}^{s-1}(g_i - 1)\theta_i + \left\{(g_s - 1)\theta_s - (-2)^{s-3}\theta_s\right\} \\
&= -(-2)^{s-3}\theta_s \\
&= -(-2)^{s-3}\min_{i\in S}\theta_i,
\end{aligned}
$$

the desired conclusion. □

Now we prove that for each two-server queueing problem, the minimal transfer rule selects the same payoff vector as the Shapley value applied to the two-server queueing game when the worth of a coalition is optimistically defined.

Theorem 10.1 *The Shapley value applied to v_O assigns each agent the same payoff as the minimal transfer rule.*

Proof Let $N = \{1, 2, \ldots, n\}$. Without loss of generality, we may assume that $\theta_1 \geq \theta_2 \geq \cdots \geq \theta_n$ and that for all $i \in N$, $d_i = i$. Note that for all $i \in N$, $g_i = \lceil \frac{d_i}{2} \rceil$. By Lemma 10.1, for all $i \in N$, the payoff assigned by the Shapley value can be calculated as follows:

$$
\begin{aligned}
Sh_i(v_O) &= \sum_{i\in S, S\subseteq N}\frac{\lambda_{v_O}(S)}{|S|} \\
&= \sum_{k=3}^{i}\frac{-(-2)^{k-3}}{k}\binom{i-1}{k-1}\theta_i + \sum_{j=i+1}^{n}\sum_{k=3}^{j}\frac{-(-2)^{k-3}}{k}\binom{j-2}{k-2}\theta_j.
\end{aligned}
$$

Let $i \in N$. By the binomial theorem and the definition of g_i, the first term in $Sh_i(v_O)$ can be rewritten as

$$
\sum_{k=3}^{i}\frac{-(-2)^{k-3}}{k}\binom{i-1}{k-1}\theta_i = \sum_{k=3}^{i}\frac{-(-2)^{k-3}}{k}\frac{(i-1)!}{(i-k)!(k-1)!}\theta_i
$$

$$
\begin{aligned}
&= \sum_{k=3}^{i} -(-2)^{k-3} \frac{i!}{(i-k)!k!} \frac{1}{i} \theta_i \\
&= \sum_{k=3}^{i} \frac{-(-2)^{k-3}}{i} \binom{i}{k} \theta_i \\
&= \frac{1}{8i} \sum_{k=3}^{i} (-2)^k \binom{i}{k} \theta_i \\
&= \frac{1}{8i} \left\{ \sum_{k=0}^{i} (-2)^k \binom{i}{k} - (1 - 2i + 4\frac{(i-1)i}{2}) \right\} \theta_i \\
&= \frac{1}{8i} \left\{ (-1)^i - (1 - 2i + 4\frac{(i-1)i}{2}) \right\} \theta_i \\
&= \frac{1}{8i} \left\{ (-1)^i - 1 + 2i - 2i(i-1) \right\} \theta_i \\
&= \left\{ -(g_i - 1) + \frac{g_i(g_i - 1)}{i} \right\} \theta_i. \qquad (10.2)
\end{aligned}
$$

Similarly, by Eqs. (10.1) and (10.2), the second term in $Sh_i(v_O)$ can be rewritten as

$$
\begin{aligned}
&\sum_{j=i+1}^{n} \sum_{k=3}^{j} \frac{-(-2)^{k-3}}{k} \binom{j-2}{k-2} \theta_j \\
&= \sum_{j=i+1}^{n} \sum_{k=3}^{j} \frac{-(-2)^{k-3}}{k} \frac{k-1}{j-1} \binom{j-1}{k-1} \theta_j \\
&= - \sum_{j=i+1}^{n} \frac{1}{j-1} \sum_{k=3}^{j} (-2)^{k-3} \frac{k-1}{k} \binom{j-1}{k-1} \theta_j \\
&= - \sum_{j=i+1}^{n} \frac{1}{j-1} \left\{ \sum_{k=3}^{j} (-2)^{k-3} \binom{j-1}{k-1} - \sum_{k=3}^{j} \frac{(-2)^{k-3}}{k} \binom{j-1}{k-1} \right\} \theta_j \\
&= - \sum_{j=i+1}^{n} \frac{1}{j-1} \left\{ (g_j - 1) - (g_j - 1) + \frac{(g_j - j)(g_j - 1)}{j} \right\} \theta_j \\
&= - \sum_{j=i+1}^{n} \frac{1}{j-1} \frac{g_j(g_j - 1)}{j} \theta_j.
\end{aligned}
$$

Altogether, for all $i \in N$,

$$\begin{aligned} Sh_i(v_O) &= \sum_{k=3}^{i} \frac{-(-2)^{k-3}}{k} \binom{i-1}{k-1} \theta_i + \sum_{j=i+1}^{n} \sum_{k=3}^{j} \frac{-(-2)^{k-3}}{k} \binom{j-2}{k-2} \theta_j \\ &= \left\{ -(g_i - 1) + \frac{g_i(g_i - 1)}{i} \right\} \theta_i - \sum_{j=i+1}^{n} \frac{1}{j-1} \frac{g_j(g_j - 1)}{j} \theta_j \\ &= -(g_i - 1)\theta_i + \left\{ \frac{g_i(g_i - 1)}{i} \theta_i - \sum_{j=i+1}^{n} \frac{1}{j-1} \frac{g_j(g_j - 1)}{j} \theta_j \right\}, \end{aligned}$$

which is the desired expression as noted in Remark 10.1. □

Remark 10.3 The minimal transfer rule for the two-server queueing problem can be characterized by imposing axioms used in Chap. 4, but appropriately adapted for the two-server queueing problem. First, it is the only rule satisfying *efficiency, Pareto indifference, equal treatment of equals,* and *independence of preceding costs.* Second, it is the only rule satisfying *Pareto indifference,* the *identical preferences lower bound, negative cost monotonicity,* and *last-agent equal responsibility*. Third, it minimizes the sum of the absolute value of transfers among agents among all rules satisfying *Pareto indifference,* the *identical preferences lower bound,* and *last-agent equal responsibility.*

10.4 A Pessimistic Approach and the Maximal Transfer Rule

Now we investigate a pessimistic definition for the worth of a coalition as in Chun (2006), which is based on the assumption that the members in a coalition are served after the non-coalitional members. Let $S \subseteq N$. We need to take into consideration the cardinality of $N \backslash S$ because this coalition is served before S. If $|N \backslash S|$ is even, then agents in S will be served from the $(\frac{|N|-|S|}{2} + 1)$th position. If $|N \backslash S|$ is odd, the last position for $N \backslash S$ is composed of one agent from $N \backslash S$ and one agent from S. Therefore, the waiting cost of agent $i \in S$, $C_i(S)$, can be calculated as follows. For all $i \in S$,

$$C_i(S) = \begin{cases} \left\{ \frac{|N|-|S|}{2} + (g_i^S - 1) \right\} \theta_i & \text{if } |N| - |S| \text{ is even,} \\ \left\{ \frac{|N|-|S|-1}{2} + (g_i^S - 1) \right\} \theta_i & \text{if } |N| - |S| \text{ is odd and } d_i^S \text{ is odd,} \\ (\frac{|N|-|S|-1}{2} + g_i^S)\theta_i & \text{if } |N| - |S| \text{ is odd and } d_i^S \text{ is even.} \end{cases}$$

where $\theta_S = (\theta_i)_{i \in S}$, $g^S \in \mathit{Eff}(\theta_S)$, and $d^S \in D(\theta_S)$. Now, the worth of coalition S, $v^P(S)$, is defined as

$$v_P(S) = -\sum_{i \in S} C_i(S).$$

Next we introduce the maximal transfer rule for the two-server queueing problem. First, this rule chooses an efficient queue. Once again, the transfer to agent i has two parts, a compensation to her and a contribution from her. Assuming an even number of agents, the contribution from agent i is the sum of the distances in the queue between her and each of the following agents times 2 (which is the number of agents in the position) divided by the number of her followers plus one. Once the contribution from agent i is determined, it is divided equally among the agents with larger indices as a compensation. Thus, the compensation to agent i is the sum of all shares of contributions from agents with smaller indices than hers.

Note that there is a symmetry between the minimal transfer rule and the maximal transfer rule. The preceding positions are considered in calculating the compensation for the minimal transfer rule, and then the preceding agents divide the amount equally and make the contribution. On the other hand, the following positions are considered in calculating the contribution for the maximal transfer rule, and then the following agents share the contribution equally.

For all $N \in \mathcal{N}$, $\theta \in \mathcal{T}^N$, all $d \in D(\theta)$, and all $i \in N$, let m_i be the contribution from agent i, defined as

$$m_i = \begin{cases} \frac{\sum_{g_j > g_i} (g_j - g_i) \cdot 2}{n - d_i + 1} \theta_i & \text{if } n \text{ is even,} \\ \frac{\sum_{g_j > g_i, g_j < \lceil \frac{n}{2} \rceil} (g_j - g_i) \cdot 2}{n - d_i + 1} \theta_i + \frac{\lceil \frac{n}{2} \rceil - g_i}{n - d_i + 1} \theta_i & \text{if } n \text{ is odd.} \end{cases}$$

Maximal transfer rule, φ^C: For all $N \in \mathcal{N}$ and $\theta \in \mathcal{T}^N$,

$$\varphi^C(\theta) = \{(g^C, t^C) \in Z(\theta) \mid \forall d \in D(\theta) \text{ and } \forall i \in N,\ g_i^C = \lceil \tfrac{d_i}{2} \rceil \text{ and}$$
$$t_i^C = \textstyle\sum_{j \in P_i(d)} \frac{m_j}{n - d_j} - m_i\}.$$

The next example illustrates the definition of the maximal transfer rule.

Example 10.3 (The Shapley Value in the Pessimistic Approach) Let $N \equiv \{1, 2, 3, 4, 5, 6\}$ and $\theta \equiv (\theta_1, \theta_2, \theta_3, \theta_4, \theta_5, \theta_6)$ be such that $\theta_1 > \theta_2 > \theta_3 > \theta_4 > \theta_5 > \theta_6$. Now we apply the Shapley value and calculate the corresponding transfers.

$$\begin{cases} t_1^C = -\frac{1\cdot 2+2\cdot 2}{6}\theta_1 \\ t_2^C = \frac{1}{5}\theta_1 - \frac{1\cdot 2+2\cdot 2}{5}\theta_2 \\ t_3^C = \frac{1}{5}\theta_1 + \frac{1}{4}\frac{6}{5}\theta_2 - \frac{1\cdot 2}{4}\theta_3 \\ t_4^C = \frac{1}{5}\theta_1 + \frac{1}{4}\frac{6}{5}\theta_2 + \frac{1}{3}\frac{1}{2}\theta_3 - \frac{1\cdot 2}{3}\theta_4 \\ t_5^C = \frac{1}{5}\theta_1 + \frac{1}{4}\frac{6}{5}\theta_2 + \frac{1}{3}\frac{1}{2}\theta_3 + \frac{1}{2}\frac{2}{3}\theta_4 \\ t_6^C = \frac{1}{5}\theta_1 + \frac{1}{4}\frac{6}{5}\theta_2 + \frac{1}{3}\frac{1}{2}\theta_3 + \frac{1}{2}\frac{2}{3}\theta_4. \end{cases}$$

It can easily be checked that these transfers coincide with the transfers assigned by the maximal transfer rule.

We note that $m_1 = \frac{\{(2-1)+(3-1)\}\cdot 2}{6-1+1}\theta_1 = \theta_1$, and this amount is shared equally by her followers, 2, 3, 4, 5, and 6, each of them receiving $\frac{1}{5}\theta_1$. Similarly, $m_2 = \frac{\{(2-1)+(3-1)\}\cdot 2}{6-2+1}\theta_2 = \frac{6}{5}\theta_2$ is shared equally by her followers 3, 4, 5, and 6, each of them receiving $\frac{1}{4}\frac{6}{5}\theta_2$ and so on.

Now we relate the maximal transfer rule of two-server queueing problems with the Shapley value of the pessimistic two-server queueing game. First, we derive the unanimity coefficient λ_{v_P}.

Lemma 10.2 *For all $N \in \mathcal{N}$ and $\theta \in \mathcal{T}^N$, the unanimity coefficient of $S \subseteq N$ is defined as follows: if $|N|$ is even, then*

$$\lambda_{v_P}(S) = \begin{cases} -(\lceil \frac{n}{2} \rceil - 1)\theta_i & \text{if } S = \{i\}, \\ 0 & \text{if } |S| = 2, \\ (-2)^{|S|-3} \max_{i \in S} \theta_i & \text{if } |S| \geq 3, \end{cases}$$

and if $|N|$ is odd, then

$$\lambda_{v_P}(S) = \begin{cases} -(\lceil \frac{n}{2} \rceil - 1)\theta_i & \text{if } S = \{i\}, \\ \max_{i \in S} \theta_i & \text{if } |S| = 2, \\ -(-2)^{|S|-3} \max_{i \in S} \theta_i & \text{if } |S| \geq 3. \end{cases}$$

Proof Let $N = \{1, 2, \ldots, n\}$. We will only treat the case when $|N|$ is even. For each $i \in N$, if $S = \{i\}$, then $\lambda_{v_P}(S) = v_P(S) = -(\lceil \frac{n}{2} \rceil - 1)\theta_i$. For all $i, j \in N$ such that $i \neq j$, if $S = \{i, j\}$, then

$$\begin{aligned} \lambda_{v_P}(S) &= v_P(S) - \sum_{T \subset S,\, T \neq S} \lambda_{v_P}(T) \\ &= -(\lceil \frac{n}{2} \rceil - 1)(\theta_i + \theta_j) + (\lceil \frac{n}{2} \rceil - 1)\theta_i + (\lceil \frac{n}{2} \rceil - 1)\theta_j \\ &= 0. \end{aligned}$$

For all $i, j, k \in N$ such that $k > \max\{i, j\}$, if $S = \{i, j, k\}$, then

$$\begin{aligned}\lambda_{v_P}(S) &= v_P(S) - \sum_{T \subset S,\, T \neq S} \lambda_{v_P}(T)\\ &= -(\lceil \frac{n}{2} \rceil - 1)(\theta_i + \theta_j) - (\lceil \frac{n}{2} \rceil - 2)\theta_k + (\lceil \frac{n}{2} \rceil - 1)(\theta_i + \theta_j + \theta_k)\\ &= \theta_k,\end{aligned}$$

as desired.

Now, as induction hypothesis, suppose that the conclusion holds for all $S \subset N$ such that $|S| \leq s - 1$. (In particular, we have shown that the conclusion holds for $|S| - 1 \leq 2$.) We need to show that the conclusion holds for $|S| = s$. We assume without loss of generality that $S = \{1, 2, \ldots, s\}$ and that $\theta_1 \leq \theta_2 \leq \cdots \leq \theta_s$. First, we consider the case when $|S|$ is even.

$$\begin{aligned}\lambda_{v_P}(S) &= v_P(S) - \sum_{T \subset S,\, T \neq S} \lambda_{v_P}(T)\\ &= -(\lceil \frac{n}{2} \rceil - 1)(\theta_1 + \theta_2) - (\lceil \frac{n}{2} \rceil - 2)(\theta_3 + \theta_4) - \cdots - (\lceil \frac{n}{2} \rceil - \frac{s}{2})(\theta_{s-1} + \theta_s)\\ &\quad - \sum_{i=3}^{s-1} \left\{ \sum_{j=3}^{i} (-2)^{j-3} \binom{i-1}{j-1} \right\} \theta_i - \sum_{i=3}^{s-1} (-2)^{i-3} \binom{i-1}{j-1} \theta_s\\ &\quad + (\lceil \frac{n}{2} \rceil - 1)(\theta_1 + \theta_2 + \cdots + \theta_s)\\ &= \left\{ (\frac{s}{2} - 1)(\theta_s + \theta_{s-1}) + (\frac{s}{2} - 2)(\theta_{s-2} + \theta_{s-3}) + \cdots + (\theta_4 + \theta_3) \right\}\\ &\quad - \left[\sum_{i=3}^{s-1} \left\{ \sum_{j=3}^{i} (-2)^{j-3} \binom{i-1}{j-1} \right\} \theta_i \right] - \left\{ \sum_{i=3}^{s-1} (-2)^{i-3} \binom{i-1}{j-1} \theta_s \right\}.\end{aligned}$$

By the binomial theorem, these terms can be rewritten as

$$\begin{aligned}&\sum_{i=3}^{s-1} \left\{ \sum_{j=3}^{i} (-2)^{j-3} \binom{i-1}{j-1} \right\} \theta_i\\ &= \sum_{i=3}^{s-1} \frac{2i - 3 + (-1)^{i-1}}{4} \theta_i\\ &= (\frac{s}{2} - 1)\theta_{s-1} + (\frac{s}{2} - 2)(\theta_{s-2} + \theta_{s-3}) + \cdots + (\theta_4 + \theta_3)\end{aligned}$$

and

$$\sum_{i=3}^{s-1}(-2)^{i-3}\binom{s-1}{j-1}\theta_s = \left\{\frac{2s-3+(-1)^{s-1}}{4} - (-2)^{s-3}\right\}\theta_s$$
$$= \left\{(\frac{s}{2}-1) - (-2)^{s-3}\right\}\theta_s.$$

Altogether,

$$\begin{aligned}
&\lambda_{v_P}(S) \\
&= \left\{(\frac{s}{2}-1)(\theta_s+\theta_{s-1}) + (\frac{s}{2}-2)(\theta_{s-2}+\theta_{s-3}) + \cdots + (\theta_4+\theta_3)\right\} \\
&\quad - \left[\sum_{i=3}^{s-1}\left\{\sum_{j=3}^{i}(-2)^{j-3}\binom{i-1}{j-1}\right\}\theta_i\right] - \left\{\sum_{i=3}^{s-1}(-2)^{i-3}\binom{i-1}{j-1}\theta_s\right\} \\
&= \left\{(\frac{s}{2}-1)(\theta_s+\theta_{s-1}) + (\frac{s}{2}-2)(\theta_{s-2}+\theta_{s-3}) + \cdots + (\theta_4+\theta_3)\right\} \\
&\quad - \left\{(\frac{s}{2}-1)\theta_{s-1} + (\frac{s}{2}-2)(\theta_{s-2}+\theta_{s-3}) + \cdots + (\theta_4+\theta_3)\right\} \\
&\quad - \left\{(\frac{s}{2}-1) - (-2)^{s-3}\right\}\theta_s \\
&= (\frac{s}{2}-1)\theta_s - \left\{(\frac{s}{2}-1) - (-2)^{s-3}\right\}\theta_s \\
&= (-2)^{s-3}\theta_s \\
&= (-2)^{s-3}\max_{i\in S}\theta_i,
\end{aligned}$$

the desired conclusion.

The case when $|S|$ is odd can be handled in a similar way. □

Now we show that the maximal transfer rule selects the same payoff vector as the Shapley value of the two-server queueing game in which the worth of a coalition is pessimistically defined.

Theorem 10.2 *The Shapley value applied to v_P assigns each agent the same payoff as the maximal transfer rule.*

Proof Let $N = \{1, 2, \ldots, n\}$. Without loss of generality, we may assume that $\theta_1 \geq \theta_2 \geq \cdots \geq \theta_n$ and that for all $i \in N$, $d_i = i$. Note that for each $i \in N$, $g_i = \lceil \frac{d_i}{2} \rceil$. First, we consider the case when $|N|$ is even. By Lemma 10.2, for each $i \in N$, the transfer assigned to agent i by the Shapley value can be expressed as

$$t_i = -(\lceil \frac{n}{2} \rceil - 1)\theta_i + \sum_{S \ni i, S \subseteq N, |S| \geq 3} \frac{(-2)^{s-3} \max_{k \in S} \theta_k}{s} + (g_i - 1)\theta_i$$

where $|S| = s$. For all $i \in N$, t_i can be rewritten as

$$t_i = \left\{ (g_i - \lceil \frac{n}{2} \rceil) + \sum_{s=3}^{n-i+1} \frac{(-2)^{s-3}}{s} \binom{n-i}{s-1} \right\} \theta_i + \sum_{j=1}^{i-1} \left\{ \sum_{s=3}^{n-j+1} \frac{(-2)^{s-3}}{s} \binom{n-j-1}{s-2} \right\} \theta_j.$$

By the binomial theorem, the terms in the first half can be rewritten as

$$\begin{aligned}
&(g_i - \lceil \frac{n}{2} \rceil) + \sum_{s=3}^{n-i+1} \frac{(-2)^{s-3}}{s} \binom{n-i}{s-1} \\
&= (g_i - \lceil \frac{n}{2} \rceil) + \sum_{s=3}^{n-i+1} (-2)^{s-3} \frac{(n-i+1)!}{(n-i-s+1)!s!} \frac{1}{n-i+1} \\
&= (g_i - \lceil \frac{n}{2} \rceil) - \frac{1}{8(n-i+1)} \left\{ \sum_{s=3}^{n-i+1} (-2)^s \binom{n-i+1}{s} \right\} \\
&= (g_i - \lceil \frac{n}{2} \rceil) - \frac{1}{8(n-i+1)} \{ \sum_{s=0}^{n-i+1} (-2)^s \binom{n-i+1}{s} - 1 \\
&\quad +2(n-i+1) - 2(n-i+1)(n-i) \} \\
&= (g_i - \lceil \frac{n}{2} \rceil) + \frac{2(n-i+1)(n-i-1) + 1 - (-1)^{n-i+1}}{8(n-i+1)} \\
&= \begin{cases} -\frac{(n-i+2)(n-i)}{4(n-i+1)}, & \text{if } i \text{ is even,} \\ -\frac{n-i-1}{4}, & \text{if } i \text{ is odd,} \end{cases} \\
&= -\frac{\sum_{k=g_i+1}^{\lceil \frac{n}{2} \rceil} (k - g_i) \cdot 2}{n-i+1}. \qquad (10.3)
\end{aligned}$$

Also, by the binomial theorem and Eq. (10.3), the terms in the second half can be rewritten as

$$\sum_{s=3}^{n-j+1} \frac{(-2)^{s-3}}{s} \binom{n-j-1}{s-2}$$

$$= \frac{1}{n-j}\left\{(n-j)\sum_{s=3}^{n-j+1}\frac{(-2)^{s-3}}{s}\binom{n-j-1}{s-2}\right\}$$

$$= \frac{1}{n-j}\left\{\sum_{s=3}^{n-j+1}(-2)^{s-3}\frac{s-1}{n-j-s+1}\binom{n-j}{s}\right\}$$

$$= \frac{1}{n-j}\left\{\sum_{s=3}^{n-j+1}(-2)^{s-3}\frac{s-1}{n-j-s+1}\binom{n-j}{s}\right.$$

$$\left.+\sum_{s=3}^{n-j+1}\frac{(-2)^{s-3}}{s}\binom{n-j}{s-1}-\sum_{s=3}^{n-j+1}\frac{(-2)^{s-3}}{s}\binom{n-j}{s-1}\right\}$$

$$= \frac{1}{n-j}\left\{\sum_{s=3}^{n-j+1}(-2)^{s-3}\{\frac{s-1}{n-j-s+1}\binom{n-j}{s}\right.$$

$$\left.+\frac{1}{s}\binom{n-j}{s-1}\}-\sum_{s=3}^{n-j+1}\frac{(-2)^{s-3}}{s}\binom{n-j}{s-1}\right\}$$

$$= \frac{1}{n-j}\left\{\sum_{s=3}^{n-j+1}(-2)^{s-3}\binom{n-j}{s-1}-\sum_{s=3}^{n-j+1}\frac{(-2)^{s-3}}{s}\binom{n-j}{s-1}\right\}$$

$$= \frac{1}{n-j}\left\{\frac{1}{4}\{(-1)^{n-j}-1+2(n-j)\}-\sum_{s=3}^{n-j+1}\frac{(-2)^{s-3}}{s}\binom{n-j}{s-1}\right\}$$

$$= \frac{1}{n-j}\left\{g_n-g_j-\sum_{s=3}^{n-j+1}\frac{(-2)^{s-3}}{s}\binom{n-j}{s-1}\right\}$$

$$= \frac{1}{n-j}\frac{\sum_{k=g_j+1}^{\lceil\frac{n}{2}\rceil}(k-g_j)\cdot 2}{n-j+1}.$$

Altogether, for all $i \in N$,

$$t_i = \left\{(g_i-\lceil\frac{n}{2}\rceil)+\sum_{s=3}^{n-i+1}\frac{(-2)^{s-3}}{s}\binom{n-i}{s-1}\right\}\theta_i+\sum_{j=1}^{i-1}\left\{\sum_{s=3}^{n-j+1}\frac{(-2)^{s-3}}{s}\binom{n-j-1}{s-2}\right\}\theta_j$$

$$= -\frac{\sum_{k=g_i+1}^{\lceil\frac{n}{2}\rceil}(k-g_i)\cdot 2}{n-i+1}\theta_i+\sum_{j\in P_i(d)}\frac{1}{n-j}\frac{\sum_{k=g_j+1}^{\lceil\frac{n}{2}\rceil}(k-g_j)\cdot 2}{n-j+1}\theta_j,$$

which coincides with the transfer for the maximal transfer rule.

The case when $|N|$ is odd can be proved in a similar way. □

Remark 10.4 As mentioned in Remark 10.3 for the minimal transfer rule, the maximal transfer rule for the two-server queueing problem can be characterized by imposing axioms in Chap. 4 appropriately adapted to the two-server queueing problem. First, it is the only rule satisfying *efficiency, Pareto indifference, equal treatment of equals,* and *independence of following costs*. Second, it is the only rule satisfying *Pareto indifference,* the *identical preferences lower bound, positive cost monotonicity,* and *first-agent equal responsibility*. Third, it maximizes the sum of the absolute value of transfers among agents among all rules satisfying *Pareto indifference,* the *identical preferences lower bound,* and *first-agent equal responsibility.*

10.5 Concluding Remarks

In this section, we investigate other properties of rules in the current context and also discuss a possible generalization of two-server queueing problems.

10.5.1 The Core, the Prenucleolus, and the Shapley Value

As shown in Chap. 3 for one-server case, it can easily be shown that the optimistic two-server queueing game v_O is concave. Thus, the allocation assigned by the minimal transfer rule belongs to the anti-core of the game.

Also, as discussed in Chap. 3, the prenucleolus (Schmeidler 1969) and the Shapley value coincide on the class of optimistic one-server queueing games. Similarly, we can ask whether these two rules coincide for optimistic two-server queueing games. As it turns out, in the optimistic approach, they coincide on the class of two-server queueing games with four or fewer agents, but not for two-server queueing games with more than four agents. Moreover, the two-server queueing games with four agents do not satisfy the sufficient conditions of Kar et al. (2009) for the coincidence of the Shapley value and the prenucleolus, either.

10.5.2 Strategic Approach

As in the case of the one-server queueing problem (see Chap. 6 for details), the domain of two-server queueing problems is convex. Therefore, the classic result of Holmström (1979) implies that in the context of two-server queueing problems, a rule satisfies *queue-efficiency* and *strategyproofness* if and only if it is a VCG rule.[2] *Strategyproofness* requires that an agent should not have an incentive to misrepresent her waiting cost no matter what she believes other agents to be doing.

[2] The family of VCG rules is due to Vickrey (1961), Clarke (1971), and Groves (1973).

On the other hand, as shown in Mitra (2005), there is no VCG rule satisfying *budget balance* for two-server queueing problems.

10.5.3 Queueing Problems with Multiple Servers

Our analysis of the two-server queueing problem can be generalized to the queueing problem with an arbitrary number of parallel servers. The minimal transfer rule for an ℓ-server problem assigns to each agent the utility in the following way: for all $N \in \mathcal{N}$, all $q = (\theta; \ell)$, all $d \in D(q)$, and all $i \in N$,

$$u(g_i^M, t_i^M; \theta_i) = -(g_i^M - 1)\theta_i + \frac{\sum_{g_j^M < g_i^M} g_j^M \cdot \ell}{d_i} \cdot \theta_i - \sum_{k \in F_i(d)} \left(\frac{1}{d_k - 1} \cdot \frac{\sum_{g_j^M < g_k^M} g_j^M \cdot \ell}{d_k} \cdot \theta_k\right),$$

where $g_i^M = \lceil \frac{d_i}{\ell} \rceil$ is the smallest integer larger than or equal to $\frac{d_i}{\ell}$. However, due to computational difficulties, it remains an open question to show the coincidence of this rule with the Shapley value.

References

Chun, Y. (2006). A pessimistic approach to the queueing problem. *Mathematical Social Sciences, 51*, 171–181.

Chun, Y., & Heo, E. J. (2008). Queueing problems with two parallel servers. *International Journal of Economic Theory, 4*, 299–315.

Clarke, E. H. (1971). Multi-part pricing of public goods. *Public Choice, 11*, 17–33.

Groves, T. (1973). Incentives in teams. *Econometrica, 41*, 617–631.

Holmström. B. (1979). Groves' schemes on restricted domains. *Econometrica, 47*, 1137–1144.

Kar, A., Mitra, M., & Mutuswami, S. (2009). On the coincidence of the prenucleolus and the Shapley value. *Mathematical Social Sciences, 57*, 16–25.

Maniquet, F. (2003). A characterization of the Shapley value in queueing problems. *Journal of Economic Theory, 109*, 90–103.

Mitra, M. (2005). Incomplete information and multiple machine queueing problems. *European Journal of Operational Research, 165*, 251–266.

Schmeidler, D. (1969). The nucleolus of a characteristic function game. *SIAM Journal on Applied Mathematics, 17*, 1163–1170.

Shapley, L. S. (1953). A value for n-person games. In H. W. Kuhn & A. W. Tucker (Eds.), *Contributions to the theory of games II*. Annals of mathematics studies (Vol. 28, pp. 307–317). Princeton: Princeton University Press.

Vickrey, W. (1961). Counterspeculation, auctions and competitive sealed tenders. *Journal of Finance, 16*, 8–37.

MIX
Papier aus verantwortungsvollen Quellen
Paper from responsible sources
FSC® C105338

If you have any concerns about our products,
you can contact us on
ProductSafety@springernature.com

In case Publisher is established outside the EU,
the EU authorized representative is:
Springer Nature Customer Service Center GmbH
Europaplatz 3, 69115 Heidelberg, Germany

Printed by Libri Plureos GmbH
in Hamburg, Germany